Daniel M. Davis:

...patibility Gene
...tiful Cure

The Secret Body

·y

HOW THE NEW SCIENCE OF THE
HUMAN BODY IS CHANGING
THE WAY WE LIVE

Daniel M. Davis

PRINCETON UNIVERSITY PRESS
PRINCETON & OXFORD

Library of Congress Control Number 2021931793
ISBN 978-0-691-21058-2
ISBN (e-book) 978-0-691-23048-1

Typeset in 12.5/15 pt Dante MT Std by Integra Software Services Pvt. Ltd,
Pondicherry

Printed on acid-free paper. ∞

Printed in the United States of America

10 9 8 7 6 5 4 3 2 1

To Katie

Contents

Come with me
On a journey under the skin
We will look together
For the Pan within.

The Waterboys

A Note to Professional Scientists

Human biology is a vast realm of science. None of it – the journey, the knowledge or its implications – is simple. I can only apologise to anyone whose work I have not included or mentioned all too briefly. Every discovery involves many students, postdocs, colleagues and collaborators, and at some level every scientific achievement is owed to a community. I apologise especially to anyone who played a role in the work I discuss here, but have not named. Through interviews with many scientists and my own reading of the original research I have sought to describe how advancements were made, but any one book can only tell part of a story. For that, I apologise in advance too. Finally, I have changed a few details in the medical stories I present in order to conceal some people's identities, but everything else of those stories is accurate and true.

Introduction

Imagine yourself as an alien with an exceptionally powerful telescope trying to understand what happens on Earth. You come across a soccer match, but your telescope isn't powerful enough to see the ball. You can make out a pitch with goals at each end, and players moving about, seemingly with some sort of organisation, but it's hard to understand what is happening precisely. You publish the observation in the *Alien Journal of Earth Science*. A few other aliens email you congratulations, but only a few.

In time, alien telescopes improve, and then occasionally you see one of the players in front of one of the goals fall over. Sometimes this is followed by the crowds of people around the pitch waving and cheering. It still doesn't make much sense, but leads to discussion at the bar during the Alien Congress of Earth Science, and your research funding is renewed. Eventually, when you are much older, a younger alien working with you notices something especially intriguing. When the player in front of the goal falls over, whether or not the crowd cheers seems to depend on one thing: whether or not the net bulges outwards. This leads your younger colleague to have a brilliant idea.

While others might have dismissed the observation without thinking very deeply about it, she wonders if there might be something there which causes the net to bulge – a ball – but it's just too small to see. At first you don't believe her, but the idea grows on you. With a ball, everything else starts to make sense: the movements of the players, the net, the cheers, the whole game, and in time other aliens agree, there has to be a ball there. Even though nobody can see the ball directly, everyone agrees it's there because so many things make sense if it is. You, your colleague and the alien who invented the super-powerful telescope collect many prizes, and everyone wants to be your friend.

Alien telescopes might improve again so that the ball is eventually seen. But equally, this might not happen. A heavy weight of evidence suggests the ball is there, but there may be no direct proof. At some level, it's debatable whether anything can ever be proven absolutely: there is no way of proving the sun will rise again tomorrow, just a heavy weight of evidence that says it will.

This tale of aliens and sport reflects how many discoveries are made. Take, for example, the discovery of the planet Neptune, first seen in 1846. The movement of another planet, Uranus, had been carefully tracked, and mathematical calculations showed that it didn't quite follow a simple orbit around the sun. This could be explained if an unseen planet was pulling on Uranus to influence its path. British and French astronomers calculated where such a planet would have to be located if it were to account for the distortion in the movement of Uranus. Then, with a telescope pointed precisely at the predicted place, the new planet was seen – Neptune. Today, a substance called dark matter and a force called dark energy are predicted to exist in order to explain the movement of stars and galaxies. As yet, both remain unseen.

Throughout almost all of history, most wonders of the human body have been hidden from view and barely imaginable. Some of our inner anatomy – bones, muscles and a few major organs – has always been available to scrutiny (albeit with a bit of delving beneath the skin), but the vast majority of our body's secrets have, until relatively recently, been the stuff of hypothesis and speculation. The discovery of cells made possible by the invention of the microscope in the late seventeenth century presaged the beginning of our modern understanding of human biology, and the discovery of the structure of DNA in the middle of the twentieth century was another gargantuan step forwards as it revealed how genetic information is stored and replicated.

Most recently, however, a whole series of technological and scientific revolutions have taken place that are revealing hidden landscapes within the human body as never before – confirming some hypotheses, undermining others and, above all, leading to a whole new realm of possibilities, both theoretical and practical.

What we are learning is that the human body is a world full of other worlds. Every organ is a menagerie of cells, and each cell has its own inner cityscape of scaffolds, capsules and monorails, all fabricated from a bewildering array of biological building materials: proteins, sugars, fats and other chemicals. Our raw materials are nothing special – oxygen, carbon, hydrogen and a sprinkling of other elements – but, put together in an exceptional way, these raw elements create a body that is conscious, self-healing and capable of poetry. We know of nothing else quite like us in the universe; there may *be* nothing else like us in the universe. Surely nothing can be more profound or enlightening than understanding how we work. And new instruments and tools, from microscopes to complex data analytics, are providing this understanding by peeling back layers of the body like never before.

Of course, all science has an ever-increasing impact on our lives, but nothing affects us as deeply or as directly as new revelations about the human body. There are any number of examples: analysis of our genes presents a new understanding of our individuality; the actions of brain cells give clues to how memories are stored; new structures found inside our cells lead to new ideas for medicine; molecules found to circulate in our blood change our view of mental health.

This book explores the recent breakthroughs in human biology that, I will argue, are vital to our future. Any number of frontiers can be considered important, but I will consider six which are unquestionably thrilling and especially impactful: the individual cell, the embryo, the body's organs and systems, the brain, the microbiome and the genome. Some of these topics you may have encountered before. If so, I hope to show how new details have recently come to light that are radically changing our understanding and capability. Other topics you may not have heard of, but are every bit as vital and game-changing as the ones that grab newspaper headlines. And at each frontier, I will show how new discoveries look set to change, or have already changed, our day-to-day lives, not to mention our overarching sensibilities and aspirations. By gathering them together in this way, I want to show that we are at the dawn of an enormous, sweeping sea change in how we live our lives. It is not self-driving cars or robots that are going to have the biggest impact on us in the foreseeable future: it's new human biology.

More than this, what is occurring in the study of human biology is reminiscent of the revolution that took place in physics during the late nineteenth century. In 1887, the German scientist Heinrich Hertz found a way to produce 'mysterious electro-magnetic waves that we cannot see with the naked eye'. Consistent with a theory developed earlier by James Clark Maxwell, Hertz showed that light is merely one type of elec-tromagnetic wave, and there are others which we cannot see, which we now know include X-rays and radio waves. At the time, it was far from clear what the practical implications of this might be – or even if there were any. Hertz died in 1894, aged thirty-six. He could not have envisaged that his work would eventually lead to the radio, the TV and the Internet. Likewise, discoveries being made about the human body now are going to impact us, our children and grandchildren, in more ways than we can even imagine.

This book is also about *how* science reveals the body's secrets, in behind-the-scenes stories of people and technology driving everything forward. As we saw for the aliens, improvements in telescopes were vital for the discovery of the soccer ball. Likewise, disruptions and advances in the prevalent under-standing of the human body are often brought about by the development of new technology. New scientific tools and instru-ments affect our lives in quieter but no less profound ways than mobile phones and social media.

Using a simple microscope in 1665, Robert Hooke saw minuscule compartments within slivers of cork, which he called cells. With today's microscopes, we can see cells shoot out protrusions, nets and packets of molecules; we see how they crawl about within our organs and tissues; and we witness the actions of enzymes and genes as they are turned on and off within them. Today's microscopes are in fact nano-scopes, capable of revealing the human body down to a few billionths of a metre.

As well as revealing new wonders about how cells work, these discoveries radically transform our ability to manipulate the body. In my own laboratory, we have used these new kinds of microscope to watch how immune cells are able to detect cancer cells and then kill them. Watching these processes unfold at a molecular scale helps us understand how immune cells recognise cancer cells and, on the flip side, how cancer cells try to avoid being caught, all of which seeds new ideas for medicines. There are currently over 3,000 clinical trials in progress, testing new cancer medicines that work by switching on or boosting the body's immune cells. Our understanding of how different immune cells react to COVID-19, and how this varies from person to person, relies on these same tools and techniques. Indeed, if there is one realm of science moved centre-stage by the arrival of COVID-19, it is human biology. Everything discussed in this book, from understanding the immune system to the human mind, also relates to what needs to be known about this virus and the next one.

But while new microscopes reveal all manner of details and opportunities, they also lead to an overarching problem. One type of microscope may capture detail best, but it takes a long time for such an accurate image to register, so another type of microscope is best for seeing movements of molecules, though it does so with less precision. A third type of microscope, meanwhile, sacrifices precision and movement in order to take a wider view – to see, for example, a slice of an organ rather than a minuscule area inside a single cell. Meanwhile, mathematical analyses and computer simulations offer a completely different perspective on the body altogether, as do analyses of gene activity or protein levels in individual cells and so on. Trying to understand the human body in this way is like trying to appreciate the *Mona Lisa* by careful examination of her left eye, or just a fragment of her brown iris. Wondrous as that is, it is not the whole *Mona Lisa*. Even the whole *Mona Lisa* is not the *whole Mona Lisa*: the painting's meaning shifts when you learn its monetary value, or about the life of Leonardo da Vinci, or how the painting deviates from other portraits from the sixteenth century. There are countless ways to understand the *Mona Lisa*, and there are countless ways to understand ourselves.

The complexity of the human body means it can only be revealed part by part, tool by tool. Just as an expert in the taste and colour of wine will gain much by being aware of the chemistry that underlies those qualities, so each perspective on the body can potentially enhance the others. And yet every scientific tool, from microscopes to mathematics, and every aspect of the body, from the brain to the microbiome, requires such depth of expertise that this tends not to happen: we tend to study the human body in silos, each community insulated from the others by its own specialised vocabulary of symbols and acronyms necessary to communicate nuances. Research communities may be dedicated to one type of scientific tool or a specific component of the body, such as one type of cell. How different types of cell communicate with one another becomes its own specialist topic. Even simple forms of life on Earth such as an individual bacterium are now rarely studied as a whole, and the human body is manifestly much more complex. As long ago as 1890, *The Times* newspaper commented that knowledge 'had already become too vast to be manageable'. Today, nobody is an expert in the whole of anything.

Many books have examined one or other specialist topic about the human body. My hope for this book is that by bringing together six key areas of contemporary biological investigation that are normally dealt with separately, we might regain a sense of the whole body and begin to see not just what the new science shows, but what it all means.

This is hard. As knowledge has become so vast, we have had to come to terms with thinking about our own body in the same way that physicists have had to deal with light being described as waves, particles or mathematical symbols. Likewise, because the human body is more complex than words or diagrams can easily depict, almost everything in a textbook is an approximation or a fragment of the whole. The deeper we examine the body's cells, for example, the more difficult it is to establish what a cell really is. Cells can swap their genetic material, for example, or directly share their innards, and some can merge together to become super-cells. Where one cell ends and another begins becomes harder and harder to define. And if cells seem hard to define, then what looked like a simple rule – all life is made up from cells – also becomes less clear. Sometimes, greater knowledge of a part leads to a diminished understanding of the whole.

For the aliens to understand soccer, the discovery of the ball was only a starting place. There's so much more to the game: the different skills of players, the tactics they use, the offside rule, the offside trap, the penalty shoot-out, the league table, knock-out tournaments, the player transfer market, the sale of television rights, the way kids playing in a school playground are influenced by their sporting heroes, the knock-on effects of traffic jams after an important Premier League match. Everything has so much depth – soccer, the *Mona Lisa*, and especially us.

But we must try to embrace it all. Because research doesn't simply lead to ever-increasing detail in our knowledge of the body's mechanics, as might be depicted in increasingly complicated textbook diagrams. This knowledge also has a huge influence on how we think of ourselves and the narrative we give to our lives. It was once thought, for example, that the body was governed by four liquid humours – blood, yellow bile, black bile and phlegm – and that illness was a result of an imbalance of one humour over the others. The truth about disease is, of course, far more fantastical than this, but it was not until the 1860s that one of humankind's greatest discoveries, the discovery of germs, opened the way to our modern understanding. For anyone alive now it is very hard, if not impossible, to know what it *felt* like to be suffering from an imbalance of the humours, but we can be sure that people did. At one time, we interpreted someone hearing voices as relevant messages from supernatural entities or an act of sorcery; now, we tell a different story about the human brain and psychosis.

More recently, we have been discovering that even germs do not account for all illness. Cancer comes about when cells in the body lose control and multiply excessively. This leads us to an awareness of all sorts of factors that we now know also contribute to ill health: excessive exposure to sunlight, radiation, chemical carcinogens and so on, which can start cells on the road to becoming cancerous. Allergies, too, have little to do with germs. Thinking about allergies has led us to other ideas about health and disease, such as the idea that some level of childhood exposure to microbes might be important in training our immune system for health: the so-called hygiene hypothesis. Understanding these causes of different kinds of disease most obviously gives us new ideas for medicine, but it also shifts the way we feel about our body and our environment: the feeling of sunlight on our skin or of growing up on a farm has been changed by the relatively recent discovery that one can be damaging and the other might be beneficial.

The effects of science on our lives also extend far beyond illness and medicine. For example, understanding evolution led to a profound alteration in our sense of origin. The fact that we share a huge fraction of our DNA with chimpanzees, and even a fruit fly, connects us in a profound way to all life on Earth. More practically, understanding hormones shapes our attitudes to teenagers, and knowing about the effects of trauma and deprivation influences how we tackle crime. There is almost no aspect of our lives that isn't framed by science's description of what's happening deep down.

Alice (that's not her real name) lost her mother when she was five years old. Her mother had died suddenly from a heart attack. Growing up in the 1980s and 1990s, Alice was bombarded with adverts promoting all kinds of products which could supposedly help keep cholesterol low, to avoid a heart attack. Alice was already anxious that she might die young, and the adverts didn't help.

One day, a letter arrived from a hospital she had never been to. The letter discussed another relative's medical situation. This relative had recently had a heart attack, and thankfully survived. But because two heart attacks in young people within the same family are very rare, doctors studied the possible causes closely. It became apparent that the two heart attacks, and other medical issues in the family, were almost certainly related to a genetic variation. By analysing blood from Alice's relative, a specific mutation had been found. The letter asked Alice if she wanted to find out whether or not she had inherited the problem.

Making a decision was especially difficult for Alice because the scientific details were vague (and they still are). The precise level of risk caused by her family's genetics was not clear. Several different mutations within the gene in question had been found in people with heart problems, but the relative risk of each – some were bound to be more dangerous than others – was not yet clear. Despite all the uncertainty, Alice went ahead with a genetic test. A few days after giving blood at her local surgery, she phoned to get the results. All of a sudden, a huge restraint on her life evaporated; she was fine, very unlikely to be at an increased risk of suffering her mother's fate. And from this, Alice's life story suddenly shifted. Day to day, she worried less about what she should or shouldn't eat. More importantly, the way she related to her parents and her family at large changed, and what she thought about having children herself. By now she was already middle-aged. Who knows what life decisions she might have made differently had this all been known earlier?

Inevitably, this kind of situation – new science shifting the way we see our lives – will arise more and more. Right now, however, a lot of that is hidden away, only discussed in detail in research labs or at the hotel bar of a scientific conference. This book will, I hope, bring the most important of that science out into public view.

To take one example, which will be explored fully in Chapter Three, let's consider again one of the new discoveries about cells. Perhaps on the face of it, basic research about the fundamental nature of cells might seem unlikely to raise any important or difficult dilemmas for our lives or for society. But I think it will.

A nerve cell is evidently different from an immune cell, and both of these are different from a kidney cell or a heart cell. But all these types of cells – nerve, immune, heart and kidney cells – are only very coarse descriptors. A fascinating new area of cell biology builds upon the idea that, on a subtle level, every single cell has its own uniqueness, influenced by its location, its age, its state of activation, its history in the body and what other cells are interacting with it. A huge global endeavour every bit as ambitious as the Human Genome Project is now under way – the Human Cell Atlas project – in which over 10,000 scientists have come together to identify and classify all 37 trillion cells of the human body. By comparing individual cells in depth – by analysing the level to which genes are activated in them, how many copies of each protein is present in them, and so on – we can classify single cells with unprecedented detail. Some of those overseeing the project hope that by scrutinising the body's cells at this scale, we could establish a periodic table for human cells – a way of organising every cell's differences in one chart that makes sense of their variety. Whether or not this particular aim pans out, everyone agrees that the project will lead to a deeper understanding of the way tissues and organs are constructed, which cells derive from which other cells in the body, and what goes wrong in disease. Excitingly, the project has already found previously unknown cells in the human body: a new type of immune cell and a new cell in the lining of the lung.

Currently a person's health is often assessed by a blood count – a simple count of how many platelets, red or white cells are present in the blood. But soon, building on the Human Cell Atlas project and related research, we will be able to assess in great detail the types, status and history of a person's blood cells. This is especially important for white blood cells, a catch-all term for countless different types of immune cells, which we already know can vary hugely between people in their specific characteristics. At the same time, the 150-year-old technique of staining tissue biopsies to categorise disease, routinely done in a hospital histopathology lab, is likely to be replaced by a far more in-depth molecular-level profiling. Taken together, these analyses will allow us to diagnose disease states to an unprecedented degree which, it is hoped, will allow us to predict whether or not a particular medicine is likely to work or could lead to toxic side-effects.

On the face of it, this is all very good news, but the implications of scrutinising the vast diversity of the body's cells reach far beyond the medical sphere. As we learn about the composition and status of the body's cells in large numbers of people, this will inevitably establish streams of new metrics by which to measure our health. Which is where things become unsettling. Inevitably, the medical profession will be asked to define what constitutes a 'normal' range for the prevalence and properties of these cells, which must in turn lead to those of us whose cells fall outside that range being categorised as 'abnormal'. We are already familiar with the idea of the body-mass index, a value derived from a person's weight and height, being used to categorise us as underweight, normal weight, overweight or obese. With the advent of new metrics by which to define a person's state of health, a whole host of new ways will arise to categorise us as normal or abnormal. Everyone will fall short in some way if enough things are measured. There are obvious implications here for health insurance premiums and, more importantly, for our psychology: such categorisation can be deeply troubling, both for an individual person's sense of self and for society's view of human diversity.

It may be that we become blasé about metrics of well-being, but so far we haven't seen any sign that this will happen. To the contrary, many people suffer from the baggage that comes with being labelled obese, for example. Somehow, being thin has come to imply attractiveness, self-control and even a kind of superiority. As we discover more and more about what makes each of us different, it becomes increasingly difficult to see what is and isn't a useful label on a person's health. Or what warrants medical intervention and what shouldn't. Many diseases are already hard to define. A person showing a set of particular behavioural traits can lead to a diagnosis of schizophrenia or autism, for example, but there is no clear-cut delineation that allows us to assess a person's behaviour and be able to state categorically that this side of the line is normal and that side is abnormal.

Just as physicists who set out to study the nature of atoms unwittingly changed the nature of bombs, so anyone working on basic science projects about the human body is likely to change society, whether they intend to or not. Which is not to say that the research should stop, or that scientists involved in this endeavour, myself included, are directly involved in something destructive; rather that, as big new concepts are opening up about how the human body works, the implications are huge, potentially explosive, and will continue to be so for some time to come.

In this book, I want us to take stock of where we're at: to immerse ourselves in the splendour of it all and to understand how we've achieved all we know – but also to think deeply about what all these new discoveries mean for our lives. I will not be afraid to speculate where they might lead and, where necessary, to challenge their direction.

In other realms of scientific endeavour, we are gathering dazzling images from space, sinking underwater drones deep into the ocean, digging up our Earth's history and pre-history, and trying to fathom the internal workings of human constructions such as our financial, social and political systems, but I think where progress is happening fastest, and where discoveries are especially likely to have far-reaching and fundamental consequences for our lives, is here, in the new science of the human body. Already we can understand and manipulate ourselves in ways that, only a few decades ago, no one could have dreamed. With so many new discoveries on the horizon, today's science fiction may one day seem naïve and simplistic in comparison to the reality that transpires.

1 Super-resolution Cells

Seeing comes before words. The child looks and recognises before it can speak.

John Berger, *Ways of Seeing*

In 1665 the English scientist Robert Hooke, then aged thirty, published the world's first picture book of microscopy, *Micrographia*.[1] The London diarist Samuel Pepys called it 'the most ingenious' book he had ever read.[2] In it, Hooke presented detailed drawings of various everyday objects dramatically enlarged for the very first time, including the unexpectedly blunt end of a needle, the mountainous edge of a razor blade and a monstrous, giant-sized flea.[3] Within a thin piece of cork, the magnified image revealed boxlike structures. Hooke named them 'cells' because they reminded him of the spartan sleeping quarters of a monastery. Three years later, the Dutch textile merchant Antonie van Leeuwenhoek probably saw Hooke's book while visiting London and went on to build microscopes himself which turned out to be better than Hooke's. In 1676, Leeuwenhoek saw tiny organisms lurking in a droplet of water: the first sighting of bacteria. A year later, by closely examining his own ejaculate, he made another crucial discovery: sperm.[4]

Then as now, microscopes uncover worlds we simply had no idea about before. And so it follows that improving microscopes – to expose ever finer details – is a sure-fire route to new revelations. But in 1872, the German physicist Ernst Abbe showed that there was a limit to how powerful a microscope could be.[5]

It wouldn't matter how well made or how perfectly aligned the optical lenses were. Even a flawlessly assembled optical microscope, Abbe's mathematical analysis showed, could not zoom in endlessly, because of the way light spreads out and bends around small objects: a feature called diffraction. The highest resolution any microscope could achieve would be about half the wavelength of light, roughly 200 nanometres (200 x 10^{-9} metres), or about 1,000 times less than the width of a human hair.[6] It's hard for us to imagine such a minuscule distance, but all sorts of wondrous and important things are smaller than this, from the structures within a butterfly's wing that provide their iridescent colouring to the HIV virus that has killed 35 million people. Other scientific instruments allow us to detect these things, albeit with difficulty, but crucially none works with living specimens. An electron microscope, for example, requires its specimen to be bathed in chemicals and then placed in a vacuum chamber.[7] Only a light-operated microscope lets us witness processes in a living cell directly, and Abbe's law seemed an insurmountable barrier to doing so beyond a certain point. On a memorial to Abbe in Jena, Germany, where he lived and worked, his law, given in mathematical notation, is literally written in stone.

And yet now, thanks to a series of discoveries so ingenious and circumstances so unlikely that they would be dismissed as ridiculous were they not the truth, we are able to see at magnitudes at least ten times smaller than Abbe predicted possible. As a result, the discovery of new human anatomy on a minuscule scale is enjoying a global renaissance, to the extent that we have had to rethink what the fundamental unit of biology – the cell – really is.

The story of this remarkable feat begins with a Japanese scientist named Osamu Shimomura and his fascination with jellyfish.

Osamu Shimomura was 'a quiet and brilliant researcher'[8] working at Princeton University with his wife Akemi in the 1960s. Nearly every summer they travelled to Friday Harbor on the San Juan Islands, around ninety miles north of Seattle, to collect jellyfish.

> We collected jellyfish from 6 a.m. to 8 a.m., then after a quick break-
> fast we would cut rings from the jellyfish until noon. We devoted all
> afternoon to the extraction. After dinner, we again collected jellyfish
> from 7 p.m. to 9 p.m., and the catch was kept in an aquarium for the
> next day.[9]

His children Tsutomu and Sachi helped, but they weren't usually up as early as their parents.[10] Locals sometimes wondered what the family were up to with their nets and buckets, scooping up so many jellyfish; they often asked, 'How do you cook them?'

In 1955, these jellyfish had been observed to emit a green glow at the rim of their umbrella-shaped bodies.[11] Shimomura wanted to understand the biological process that made them glow. At least initially, he didn't have any practical application in mind for his work. He was simply fascinated by the way some animals glow. All kinds of life – including fireflies, worms and deep-sea fish – use light to attract mates, warn off predators and communicate in ways we hardly appreciate. Life continues to surprise us with its colour: flying squirrels have recently been found to shine pink under UV light, and nobody knows how or why.[12] Shimomura wanted to understand the basic principles of how this happens.[13]

Shimomura's success was partly owing to his characteristic approach to solving problems. Rather than foraging through books and scientific papers to find a suitable method, he would devise his own procedure from scratch with unusual resourceful-ness. Instead of using one of the filters that happened to be available from the lab supply store, for example, he would think about the kind of the fabric that would work best and seek that out, wherever it was to be found. His daughter Sachi recalls how her father would often wander around a hardware store looking for things he could repurpose in the lab. He used dental floss to sew netting onto metal wire frames to make the shallow dip-nets his family used to collect the jellyfish. His jellyfish-cutting machine was essentially made from a juice blender.[14] Shimomura would often emphasise this as an important ethos: that young scientists need to learn how to learn, and inventing one's own procedures is an important way of doing so.

This approach to science came from his upbringing. His family moved homes several times, and his father, an army captain, was away a lot. Shimomura's school education was frequently disrupted by military exercises and later abandoned entirely. At sixteen, he was at work in a factory just 15km away from Nagasaki when the atomic bomb was dropped. He witnessed two B-29 planes drop parachutes without any people hanging from them and, as he recalls in his autobiography, 'a powerful flash of light hit us through the small window. Then, maybe forty seconds after the flash, we heard a loud sound and felt a sudden change of air pressure.'[15] On his way home that day, a black rain fell. His grandmother gave him a bath as soon as he got in, which probably saved him from radiation poisoning.[16] Growing up in Japan during the Second World War taught Shimomura to be strong, independent and resourceful.[17]

Ultimately, by comparing extracts from the jellyfish cells, seeking any that showed optical activity, Shimomura identified two types of protein molecule that make jellyfish cells glow. One emits blue light in the presence of calcium and a second takes up the blue light and emits green light.[18] It was this second protein, later named green fluorescent protein or GFP, that was to play such a crucial role in microscopy.[19]

It was not until years later, though – at just after noon on Tuesday, 25 April 1989, to be precise – that Chicago-born Martin Chalfie, working at Columbia University, New York, happened to sit in on a talk which mentioned Shimomura's work, and a new chapter in the story of GFP began.[20] Immediately, Chalfie began to fantasise about how this green-glowing protein might be used inside cells of other animals – specifically a small worm that he was studying – to highlight the location of specific types of cell or even certain molecules within cells.[21] In an era before Google and Wikipedia, he spent the next day phoning people in order to find out all he could about it.[22]

One person he was led to call was Douglas Prasher, then at the Woods Hole Oceanographic Institution, who was working to identify the gene which carried the instructions for the production of GFP. Prasher agreed to send Chalfie the gene once he had isolated it, but soon afterwards they lost touch. In time, Chalfie went on a sabbatical. Unable to reach him, Prasher assumed he had left science altogether. And when Chalfie never heard from Prasher, he assumed Prasher had never isolated the gene. It was not until 1992 that Chalfie stumbled upon a scientific paper by Prasher saying that he had.[23] Chalfie got back in touch, and Prasher sent him the gene.

In Chalfie's lab, they found that the jellyfish gene could indeed be re-deployed to make bacteria or worms glow green.[24] It was a PhD student, twenty-six-year-old Ghia Euskirchen, who was the first person ever to see this. The bacteria's green glow was so faint that Chalfie's lab microscope couldn't detect it. Luckily, she double-checked on a microscope in another lab and discovered that her experiment had worked.

It was already well established that genes could be transferred between species – because the basic chemistry of genes is the same in all life on Earth – but the fact that it took only a single gene to make an organism glow green was a vital revelation: it could have easily been the case that GFP would only work in concert with a suite of other proteins that were only found in those particular jellyfish. Chalfie's lab first described these results in the October 1993 issue of *Worm Breeder's Gazette* – not a widely read publication, and certainly not a usual source for paradigm-shifting new technology.[25] 'We have lots of ideas of how GFP might be used and imagine that other people will have many more,' they wrote. 'If you are interested in obtaining [the GFP gene], please write … we'd like to know what you are interested in doing, but that's not essential.' Soon after, in February 1994, they published their work in the pre-eminent journal *Science*.[26]

Eventually, the green jellyfish protein would be used in a vast array of experiments to study all kinds of life, from yeast to humans, but when Chalfie first talked to others in his university department about it, few grasped its potential. He thinks this is probably because it's hard to realise the full importance of anything new the first time you hear about it.[27] But one person who did appreciate the work very early on – and she undoubtedly heard about it far more than once – was his wife, Tulle Hazelrigg, also a professor at Columbia. It was in Hazelrigg's lab that the major step was taken that turned GFP into such a useful device: her team attached GFP to another protein by fusing together the two genes that encoded for them, allowing scientists to 'tag' that protein with GFP and thereby detect its location inside a cell. With this, Chalfie's fantasy had come true: the green-glowing jellyfish protein had become a tool for watching life on a minuscule scale – because any particular type of protein could be tagged with GFP and watched.[28] A biological laser pointer, as *Discover* magazine called it.[29]

In 2008, Shimomura and Chalfie, along with Roger Tsien at the University of California, San Diego, who improved the brightness of GFP and developed other proteins to glow in different colours, won the Nobel Prize for Chemistry. But the Nobel committee left out Prasher – the rules of the prize allow a maximum of three winners. When he heard the news, he was working for a Toyota car dealership in Huntsville, Alabama. He had struggled to get funding for his research at the Woods Hole Oceanographic Institution, worked for a while at the US Department of Agriculture, then took a job with a NASA contractor in Huntsville. Politics then changed NASA's priorities and his project had been shut down. After a year of unemployment and bouts of depression, he had taken the job at the dealership so that he wouldn't have to move city and his daughter could stay at the same high school.[30] So while the Nobel winners were set to receive several hundred thousand dollars each in prize money, Prasher was on $10 an hour.

Chalfie and Tsien got in touch and paid for him and his wife to attend the prize ceremony in Stockholm.[31] Both mentioned his contribution in their Nobel lectures. Over a three-year period, Prasher, like Shimomura before him, had caught many tens of thousands of jellyfish.[32] He had eventually isolated the gene for GFP, which was undoubtedly a crucial step towards its use as a tool, but he didn't begrudge others winning the Nobel: 'Do I feel cheated or left out? No, not at all. I had run out of funds, and these guys showed how the protein could be used, and that was the key thing.'[33]

Nobody could possibly have known how research into jellyfish would lead to something so valuable to so many branches of biology. Scientific breakthroughs happen in all sorts of mysterious ways. Late in his life, Shimomura noted that after about 1990 the jellyfish became scarce in the waters where he used to collect them, probably because of pollution and perhaps specifically as a result of the *Exxon Valdez* oil spill off Alaska in 1989.[34] If the jellyfish had disappeared from there thirty or so years earlier, he would never have discovered GFP.

And if Shimomura had never discovered GFP, then a middle-aged, retired scientist from Michigan might never have built a groundbreaking new kind of microscope in his friend's living room.

Born in Ann Arbor, Michigan, Eric Betzig had always been driven to do something transformative: 'I grew up with Apollo and *Star Trek* and wanted to make a warp drive.'[35] After completing a doctorate at Cornell, he went on to work at Bell Labs, where the transistor and laser were invented and developed, and famous for its go-getting atmosphere.[36] Here Betzig worked on the improvement of microscopes, but after six years at Bell, he was fed up. The type of microscope he was working on seemed to him a technological dead end, and he thought others in his field were jumping to conclusions that weren't justified. He was in any case exhausted by the relentless hard slog of research. And he could sense that the phone company AT&T, which funded Bell Labs, was finding it harder and harder to justify the expenditure of so much money on basic science. In 1994, burnt out and disillusioned, he quit.

Betzig stayed at home with his daughter. Still, he was unable to get science entirely out of his system. One day, he was walking his daughter in a stroller when he hit upon an idea for a new type of microscope. He published the idea but left it at that, making no attempt to build the instrument he'd conceived.[37]

Eventually, in 1996, he went to work in his father's successful machine tool company. Here, however, he came to realise how many constraints there were on the design of new equipment in a business environment: a new machine had not only to fulfil its particular task but also be cost-effective, safe, reliable and thoroughly documented. He missed explorative science, and felt that he wasn't such a great businessman. So in 2002, he quit again. Aged forty-two, with two young kids, no job and no prospect of a job, he did not seem like a man on a path to winning a Nobel Prize.

Without a career plan, he reconnected with a friend from his Bell Labs days, Harald Hess. They met up in national parks and chatted about the meaning of life and the impact they wanted to have before they died. 'What we realised,' Betzig recalls, 'is that while neither one of us fits well in the normal academic scheme of things, we both really love science, and we love the ability to be able to pursue our curiosity.'[38] Betzig decided to catch up with the latest advances and, excited by what he read, realised: 'Oh, shit, I've got to do microscopy again.'[39]

What he'd found so inspiring – 'My jaw was hanging down for a week in astonishment at this'[40] – was how individual protein molecules could now be tracked within living cells by tagging them with GFP. This was the missing link that could allow microscopes, like the one he'd conceived of some seven years previously, to work with living cells. So, with his long-term friend Hess, he set about building it. They worked in Hess's living room because, Betzig recalls, Hess wasn't married. They used equipment that Hess had kept in storage from their time at Bell Labs and bought other parts with $25,000 each of their own money. Even though he'd been unemployed for two years and there was no guarantee the microscope would work, Betzig's wife understood that he *had* to do it.[41] 'It was an obsession … a once-in-a-lifetime opportunity.'[42] You could spend that amount of money renovating your bathroom, Hess said, but this was so much more interesting.

Their equipment sat on a plastic mat over the living-room carpet, connected to a computer propped up on a cardboard box.[43] They worked day and night, rekindling the intense work ethic of their Bell Labs days. Betzig sometimes fell asleep on the couch while Hess carried on. Outside the normal infrastructure of science, they were undistracted and ultra-focused. Their goal was nothing less than to build a microscope that would break Abbe's law – and overcome a fundamental law of physics.

Crucial to their plan was the recent invention of a version of the green-glowing jellyfish protein that could be 'switched on' when bathed in blue light.[44] Betzig and Hess's big idea was this: instead of bathing the cell continuously in blue light so that the entire cell glows green, they would submit the cell to very brief flashes of low-level blue light, so that only a few of the GFP molecules would be switched on to glow at the same time. Chance dictated that the few molecules made to glow each time would be sufficiently far apart from one another that each one would appear as an isolated dot of light. In accordance with Abbe's law, the image of each glowing molecule would be blurred, but the exact position of each could be inferred to be right at the centre of its spot of light. Through repeated exposures, with a different selection of molecules being switched on randomly each time, gradually the co-ordinates of every molecule that had been tagged with GFP could be discovered. When reassembled into a single picture using computer software, an image of all the tagged molecules within the cell would be arrived at with far greater resolution than could have been achieved otherwise.[45]

Things moved fast in Hess's living room, and soon they had a prototype complete. But to test their microscope with living cells they would need the help of a biologist. Betzig was due to give a talk at the National Institutes of Health on a different topic, but he knew there would be one scientist there, Jennifer Lippincott-Schwartz, who might be open to helping him. Her career had been based on using new kinds of microscopy to study cells, and one of the scientists in her team, George Patterson, had developed the version of GFP that Hess's and Betzig's technique relied on.[46] On the morning of Betzig's talk, Lippincott-Schwartz remembers someone phoning and asking if she would come to the seminar because the speaker, Betzig, had requested she attend. She hadn't planned on it but agreed to do so and, since it sounded like the talk would be about microscopy, she asked Patterson to join her. Having changed her plans to hear Betzig, she said to Patterson as they were walking over, 'This guy better be pretty good.'[47]

Within five minutes of Betzig beginning his talk, she realised he wasn't just good, he 'was a whole different level of scientist'.[48] After the talk, over lunch, Betzig pitched the new microscope to them. Lippincott-Schwartz and Patterson could easily have dismissed Betzig and Hess as two crazy guys who hadn't published a scientific paper in over ten years. But to their credit – and also testament to Betzig's confidence and charisma – Lippincott-Schwartz and Patterson were enthusiastic. With their go-ahead, Betzig and Hess packed up and rebuilt their microscope in a room at the National Institutes of Health which, as Betzig recalls, was a lot less comfortable than Hess's living room. It worked. They quickly found that they could locate molecules in living cells with unprecedented accuracy.[49]

It took six months from the moment they started building the instrument to proving it worked and getting enough data to earn a Nobel Prize. 'We knew we had to work fast because this idea was going to be ripe and in the air,' Betzig recalls.[50] They were right to hurry. Xiaowei Zhuang at Harvard University – who had been educated in China in a special programme for gifted children – developed a very similar type of microscopy, except that she used a chemical dye, rather than the jellyfish protein, as a label.[51] Zhuang demonstrated how well her microscope worked by looking at dyes along strands of isolated DNA. Her work was formally published one day before Betzig and Hess's.[52] A third team, at the University of Maine, also developed a similar microscope.[53]

Stefan Hell, working at the Max Planck Institute for Biophysical Chemistry in Göttingen, Germany, also developed a new kind of microscope that smashed Abbe's law, but his method was completely different. Born and raised in Romania, Hell moved with his family to Germany in 1978, when he was aged fifteen. As an only child, he spent a lot of time with books, enjoyed science fiction thrillers on TV, and knew from a young age that he wanted to be a scientist. He later wrote that while he was growing up in Communist Romania, a feeling took root which would prove prescient: 'Things which are publicly asserted and constantly repeated aren't necessarily true.'[54]

Hell was attracted to working in theoretical physics, but because his parents struggled when they moved to Germany – his father's job was uncertain and his mother was diagnosed with a serious illness – he thought he should work on something more vocational. So for his doctoral research, he worked in a small start-up company developing microscopes to help with the production of computer chips. The work was practical, as Hell had wanted, but also boring. He felt that the physics of microscopes was the physics of the nineteenth century. He was trapped between the need to earn a living and the desire to work on something scientifically challenging. Seeking a way out, he wondered whether there might not be something left to do in microscopy that could still be important. His thoughts turned to Abbe's law, and he began to question if it was truly irrefutable.

The critical problem which limits a microscope is that a lens cannot concentrate light beyond a certain point because of light's wavelength. If two molecules lie *within* the spot where the light is focused, they will both be illuminated and there's no way for the microscope to distinguish them. Hell knew that nothing could solve the problem directly, but he thought there had to be some trick that could circumvent the problem, just as when a helicopter flies, it doesn't alter the fundamental physics of gravity, but defies it using the trick of rapidly rotating blades to gain lift. He thought about the problem for many years and studied countless textbooks and scientific papers, searching for something that might work.[55] Eventually he had a brilliant idea: rather than reducing the size of the beam of light that shines on the specimen, perhaps he could change the area from which the light was then detected.

Something he read in a quantum optics book while he was working in a lab at the University of Turku in Finland gave him the vital clue he needed: molecules that have the ability to fluoresce, like the green-glowing jellyfish protein, can also be *prevented* from doing so by shining light on them of a specific frequency. From here, his brilliant idea followed. He would build a microscope that used two laser beams aligned to hit the sample in exactly the same place, with the second laser beam having two vital properties: it would be tuned to switch *off* the fluorescent molecules, and instead of being a normal beam of light, it would be a tube, resulting in a doughnut- or ring-shaped spot.[56] So, as the first laser illuminated a spot on the cell, switching the molecules within it *on*, the second would switch *off* those molecules around the spot's outer edge, meaning light would be emitted only from a central bullseye, one smaller than Abbe's law would otherwise allow.[57]

Hell first published this idea in 1994 under the name Stimulated Emission Depletion, or STED, microscopy.[58] In 1999, he and his team built the microscope and showed it worked.[59] The world's most prestigious scientific journals, *Nature* and *Science*, both refused to publish Hell's paper, on the grounds that this result didn't reveal any new biology and would therefore, they said, be of limited interest.[60] They couldn't have been more wrong.

In 2014, Betzig and Hell won a Nobel Prize, alongside William Moerner from Stanford University, who studied the properties of GFP and had been the first person to optically detect a single molecule.[61] But because, again, the Nobel Prize can be awarded to a maximum of three people, Betzig's friend Hess was left out.[62] In his acceptance speech, Betzig said: 'One of the bittersweet things about winning this award is not having [Hess] here by my side up on the stage.'[63] Hess is like an angel, Betzig said to me in 2019; 'He was happy for me to win, and if it was the other way round, I'm not sure I could be so generous.'[64]

The Nobel Prize they won was for Chemistry, which is perhaps surprising, given that none of these pioneers were chemists. The properties of molecules – their chemistry, if you like – underlies the way these new microscopes work, but really the triumph here encompasses physics, chemistry and biology, not to mention computer science, maths and electronics, which were all vital too. As Betzig said to the *New York Times* in 2015:

> You know, I'm not comfortable with labels. I'm trained in physics but don't think of myself as a physicist. I have a Nobel Prize in Chemistry, but I certainly don't know any chemistry. I work all the time with biologists, but any biology I have is skin-deep. If there is one way I characterize myself, it's as an inventor.[65]

Having a PhD in physics myself, and now working in human biology, I agree with Betzig that labels are often not useful. But it's something else Betzig said that I find especially inspiring, and often think about. Many Nobel lectures end with the winner thanking all the people who helped them along the way. Betzig did the same, and especially thanked Hess, but then he went on:

> The last thing I would like to say … is about taking risks. People are always exhorted to take risks, and that's fine, but you're hearing that from guys whose risks paid off. It's not a risk unless you fail most of the time. And so what I'd really like to do is I'd like to dedicate my talk to all of the unknown people out there in any walk of life who have gambled their fortunes, their careers and their reputations to try to take a risk, but in the end, failed. I'd just like to say that they should remember that it's the struggle itself that is its own reward, and the satisfaction that you knew that you gave everything you had to make the world a better place.[66]

I sit in a darkened room with the temperature exquisitely controlled – there must be no flux in the environment. The machine itself fills two large tables. The main body of the microscope sits on a table that is especially bulky, because it includes a pneumatic system to isolate it from ambient vibrations in the room. To witness nature on a nanoscale, things need to be held steady on a nanoscale. A series of metal boxes, stacked to the side of the tables, house lasers and their electronic controls, which feed light into the microscope along optical fibres. I rarely need to look down the microscope's binocular eyepiece because what I would see is also displayed on a large computer screen in front of me. An adjacent screen shows graphics of sliders and drop-down menus to adjust the power of the lasers, the sensitivity of the light detectors, the pixel size, the speed at which the lasers scan the sample, the number of times a sample is scanned, the distance the objective lens moves to capture different depths, the pinhole size, and much more. The software is marketed as intuitive, but it takes some getting used to. You have to bond with the hardware too, tweaking the position of the various elements to get everything perfectly in tune, just as I imagine an electronic musician does with their audio samplers and synthesisers to get the perfect sound. For anyone who hasn't used a super-resolution microscope before, the experience is other-worldly. Going for a walk in a field, in a forest or along a trail, brings us close to nature, but in a blackened room with the air hardly moving, we witness its deepest secrets.

Super-resolution images obtained in my lab have led to a new idea for treating patients with a rare genetic disease called Chediak-Higashi syndrome. Children with this syndrome are unable to fight infections that would normally be dealt with easily, and often die young. In normal circumstances, immune cells kill aberrant cells – including cancer cells or virus-infected cells – by secreting toxic enzymes into them. These enzymes are stored inside immune cells within small droplets of liquid, called lytic granules, each enclosed by a thin layer of fat molecules. When an immune cell encounters a diseased cell, such as a cancer cell or a virus-infected cell, receptor proteins protruding from the surface of the immune cell will detect molecules on the outer coating of the diseased cell that identify it as a threat. The immune cell will then flatten up against the diseased cell, establishing a tight surface contact. Once the cell is in position, the lytic granules – containing the toxic enzymes – take about a minute to gather together at the edge of the immune cell, next to the diseased cell, and there they pause momentarily. Then, in a process that still isn't entirely understood, some of these lytic granules fuse with the outer edge of the immune cell (the coating of the lytic granules and the surface of the whole cell are made up of similar fat molecules), so that their contents – the deadly enzymes – are expelled from the immune cell onto the diseased cell. In a few minutes or so, the diseased cell visibly bulges and bubbles. Less easy to see directly, the diseased cells' proteins and genetic material are chopped up and degraded. Remnants of the dead cell are then engulfed by another type of immune cell, where they will be broken down further and their chemical components re-used, in the same way that when we are buried, our molecular parts may be re-used by organisms in the earth.

But in children with Chediak-Higashi syndrome this process doesn't work. Working with Polish scientist Konrad Krzewski at the US National Institute of Allergy and Infectious Disease in Bethesda, therefore, we deliberately mutated a gene known to cause Chediak-Higashi syndrome in immune cells in a lab dish, and examined them with a super-resolution microscope. We hoped to understand how this genetic mutation changed immune cells, to help explain why children with this syndrome are especially susceptible to certain types of infection.

We found that these genetically altered immune cells had larger-than-usual bags of toxic enzymes inside – about twice as big as normal. We discovered that they were simply too big to pass through the structural meshwork – a bit like the strings of a tennis racquet – that underlies the cell surface and gives the cell its shape, and would therefore be unable to launch an attack on diseased cells.[67] This could indeed be part of the reason why children with this syndrome can't deal with some types of infection very well, because their immune cells can't easily launch an appropriate attack.

This in turn led us to think that finding a way to open up the meshwork – increase the size of the holes between the racquet strings – might restore the affected immune cell's ability to kill diseased cells.[68] I knew about a drug that can do precisely this, used to treat patients with certain types of cancer, because it kept my own father alive. It is also responsible for one of the world's worst ever medical tragedies.

The use of thalidomide to help pregnant women with morning sickness led to many thousands of babies being born without fully developed limbs and with a host of other deformities. Roughly half of them subsequently died young. Nobody knows how many miscarriages were caused by the drug. Compassionate and conscientious doctors had to care for 'thalidomide children', as they became known, aware that they themselves had suggested their mothers use the drug, thinking it would help.[69] However, thalidomide was also observed to have some positive effects on various diseases, including leprosy and cancer. The US pharma company Celgene created a safer derivative of thalidomide, sold as Revlimid, by switching one oxygen atom for a nitrogen atom in its chemical structure. My father, afflicted with multiple myeloma, took this drug for many years. It's not entirely understood how it works – thalidomide and its derivatives have many effects in the body – but one thing it does do, as we found out in my own lab, is boost the opening up of an immune cell's structural mesh, making it easier for them to kill cancer cells.[70]

Krzewski and I first got chatting about Chediak-Higashi syndrome at the hotel bar during a scientific meeting in Heidelberg, Germany, in September 2013: the most valuable encounters at scientific meetings are usually the informal ones. He was studying the illness directly and my lab had expertise in using super-resolution microscopy to watch immune cells kill. Although we didn't have any clear plan at first, it seemed like we should join forces. I had a Polish researcher in my lab at the time, Ania Oszmiana, who was also at that meeting. That she and Krzewski shared a language and culture probably helped get things going – rapport between scientists is at least as important as a good idea. Eventually, this led us to test whether the drug my father was taking to treat his cancer might also help children with Chediak-Higashi syndrome. By the time we arrived at a clear set of experiments to do, Ania Oszmiana had achieved her doctorate, largely based on other work using super-resolution microscopy, and she had left my lab to work in Australia. These experiments were done by an Ethiopian student in my team, Mezida Saeed.[71]

Giving children with the syndrome the drug directly was not an option, and, besides, we couldn't then have given them a deliberate viral infection to see how they fared. Instead, we isolated immune cells from their blood and tested whether adding the thalidomide derivative would rescue their ability to kill diseased cells in a lab dish. The answer turned out to be yes, to some extent. This is not a medical breakthrough, because we didn't try any experiments on animals or humans, and the drug could, for example, have unwanted side-effects. But scientifically, it was a useful advance – understanding a disease and what sort of approach might work as a treatment – and all brought about by super-resolution microscopy.

In my view, there are two ways to use a super-resolution microscope. Most commonly, it is used in the way I have just described: to investigate a process we already know to be important – in this case, how toxic proteins emerge from an immune cell to kill a diseased cell – revealing crucial new detail. But the other way to use a super-resolution microscope is more akin to the way Hooke used a microscope in 1665: to explore nature, without setting out to see anything in particular. By using a super-resolution microscope simply to watch cells or combinations of cells, something entirely new might be revealed. Perhaps a new part of a cell will be discovered, or an unexpected way in which two cells interact will be witnessed. Both approaches – digging into the details of known mechanisms and open-ended exploration – are vital. But it's the second approach that leads to the most magical feeling of discovery.

Jennifer Lippincott-Schwartz and her lab team were the first cell biologists to use Betzig and Hess's microscope after they relocated it from Hess's living room to her lab. She has spent her whole career using new technology to understand cells, and is very familiar with what often happens when you see something unexpected for the first time: others don't believe you.

Having earned a degree in philosophy and psychology, she has given a great deal of thought to why people react in this way – what it takes to change someone's view of the world and the sociology of science. Understanding that it takes time for a scientific community to come to agreement about anything new is what has given her strength to persevere in the face of criticism.[72]

At a prestigious scientific meeting in 1998, before super-resolution microscopes had been built, Lippincott-Schwartz presented a movie, itself a novelty at the time, made up of a series of microscope images captured at intervals of a couple of seconds, revealing how protein molecules move from one particular place to another within a cell.[73] Previously it was thought, based on indirect evidence, that small bags or droplets known as vesicles carried these proteins, shuttling them from place to place, but her movies revealed direct evidence of something else: tubular structures were ferrying the proteins, with no small vesicles to be seen. Rather than take her movies at face value, though, someone in the audience asked: where are the small vesicles? Someone else in the audience suggested that they were invisible because her microscope simply couldn't detect them. Needless to say, Lippincott-Schwartz was proved right – there were no invisible vesicles, as lots of methods eventually showed – but it took some time for the community to shift its thinking. When asked about the drive it takes to persevere with ground-breaking research, she says, 'I don't like doing things which are not significant.'[74]

In 2016, Lippincott-Schwartz's team used a super-resolution microscope to look at the elaborate structure inside cells where proteins are manufactured and processed, called the endoplasmic reticulum, or ER.[75] It was thought that this structure, which fills a large part of the cell, was made up of sheets and tubes of membrane. But it turns out that this view, found in high school textbooks, wasn't really right either. Lippincott-Schwartz's team revealed that the supposed sheets of membrane were in fact tubular structures, too, so densely packed that when viewed under a normal microscope they looked like flat sheets of membrane. There had been nothing to suggest that this would be the case. It was an entirely unexpected discovery. Super-resolution microscopy has set us a new challenge: understanding what this means. A dense tubular structure might increase flexibility, which could be important when the cell moves. Or it could provide greater surface area, the better to store or facilitate reactions. As yet, we do not know.

In a similar way, another new structure has been discovered inside axons, the long, slender projections that connect our nerve cells to other cells. Zhuang, who had developed super-resolution microscopy around the same time as Betzig, used the new technology to discover a series of protein rings lining the surface of axons.[76] The rings are so close together that they can't be seen when viewed through a normal microscope, which is why they hadn't been detected before.[77] This structure, named the membrane-associated periodic skeleton, has now been seen in axons protruding from every type of neuron that has since been examined, including neurons from a wide range of animals.[78] Once again, nobody predicted its existence, and we now need to understand what it's for. Perhaps it gives axons strength that is essential for their survival throughout a person's life. Or it might play a role in the transmission of electrical impulses along the axon's length in some way that we don't yet understand.[79]

Exploring cells with these new microscopes is akin to the moment you put on a new pair of prescription glasses. Details are revealed which you had no idea were there. The technology is still so new that a tremendous amount is still being discovered. One of the most tantalising discoveries – with evidence accumulating from new microscopes as well as other technologies – is that cells send out small bags of genetic material and proteins as a means of communication with other cells. As far back as 1983, it was shown that membrane-enclosed vesicles were jettisoned from cells.[80] But at first, most scientists thought that these were small bags of trash, carrying away biological components that the cell no longer needed. In 1996, however, it was discovered that vesicles had the ability to alert immune cells to the presence of a problem, such as a virus infection.[81] Then, in 2007, a team based at the University of Gothenburg, Sweden, showed that vesicles also carry genetic material.[82] The implication is that cells can send out exquisitely complex messages – in the form of bundles of protein and genetic material – to other cells. Tools and information can be shared perhaps to help establish integrated communities of cells in our organs and tissues.[83] Philosophically, such complex integration between cells can be considered a challenge to the central doctrine of what a cell is. The individuality of one cell versus another is less clear if any number of their components can be physically shared.

Leaving aside this debate, for which there's no easy answer, we now know that there are at least two types of vesicles that cells can emit. One type, micro-vesicles, form like buds at a cell's surface, while another type, exosomes, are assembled inside the cell. These are only broad descriptors, however. In the same way that we refer to immune cells and nerve cells when in fact there are many different types of immune cell and nerve cell in the body, there are no doubt many different varieties of vesicles within these two categories. The menagerie of small vesicles released from cells – and what they do in the body – is still being explored. Some are likely to be long-lived and circulate in the blood to impact distant organs or tissues, while others probably break down and release their contents locally. In at least one situation, vesicles may even move between people.

Astonishingly, human breast milk contains vesicles that encapsulate nearly 2,000 different proteins.[84] Some of these proteins have been studied in other situations and found to regulate cell growth and influence the immune system. This leads to the idea that vesicles in breast milk might aid the development of a baby's gut and immune system. But before this affects anyone's decision about breastfeeding or using formula milk, it is crucial to note that this is only an idea, and one that is extremely hard to test directly; we are at the cutting edge of knowledge here and much is unknown.

Vesicles may also play a role in disease. There is evidence, for example, that vesicles can contribute to a build-up of fatty deposits (called plaque) in arteries, which can in turn cause life-threatening problems such as heart attack or stroke.[85] Other types of vesicles might be crucial to how cancer spreads in the body. Vesicles from a primary tumour, in a person's breast for example, can enter the bloodstream and land somewhere else in the body, such as a person's lung or liver, where they then unload their cargo, preparing this new location for the arrival of the cancer.[86] It's conceivable, then, that new medicines could work by blocking the genesis, movement or activity of vesicles. In the short-term, however, vesicles are most likely to prove useful in diagnosing disease. The contents of a person's vesicles isolated from a blood sample, for example, might be used to indicate the state of a person's health, assess the type of cancer a patient has, and so on. Eventually, vesicles might also be exploited directly as a drug-delivery system. Vesicles might, for example, be constructed to deliver gene-editing tools into cells – a subject we will return to.

Cells are often said to be life's basic building blocks. But this conjures up an image of cells being like Lego bricks. Thanks to super-resolution microscopes, and other technologies, we are discovering that if a cell *is* like a Lego brick, it's one that can change its size and shape, that has the ability to move, multiply and kill off other, damaged Lego bricks, and can send out small packets of information that change the nature of Lego bricks far away. It's fair to say that Lego – or anything else manufactured by us – has nothing on life.

In the same way that Samuel Pepys stayed up until 2 a.m. reading Hooke's *Micrographia,* I relish the new views of cells described in this chapter. They reveal an intricacy to what we are, far beyond anything we might have imagined without the development of super-resolution microscopes and other tools. These details are magical and humbling. But also, personally, I find it existentially unsettling to realise how much is going on within my body without my awareness. The discoveries described in this chapter elevate that feeling to a whole new level.

This new world – the nanoscale anatomy of the human body – was opened up not by large corporations or a government strategy, but by a few rebels with a big vision, built upon by thousands of scientists around the globe, leading to new instruments that let us see ourselves more clearly than ever before. The technology continues to improve. Other new microscopes are being built right now, allowing us to see more and see better. New wonders will be found that will impact our lives, not least in creating whole new categories of medicine. The rebels planted trees that will bear fruit for decades to come.

2 The Start of Us

My birth had caused a worldwide sensation and thrown up all kinds of
moral and religious arguments.

Louise Brown, *My Life as the World's First Test-Tube Baby*

In 2006, Magdalena Zernicka-Goetz had a genetic test which
indicated her unborn baby might carry a serious abnormality:
an extra copy of chromosome 2. This chromosome holds
around 8 per cent of the human genome. An additional copy
of so many genes can have any number of effects on a baby's
health and development, including an increased chance of
miscarriage late in pregnancy. Crucially, the test indicated that
not all of the baby's cells were likely to have this extra copy of
chromosome 2; about a quarter of the cells taken from the
placenta, which is derived from the foetus, showed the abnormality. 'As a woman, I wanted to believe there was hope,' she
recalls. 'As a scientist, my instincts from my work in this field
told me that there could be.'[1]

Born and raised in Warsaw, Poland, Zernicka-Goetz had once
dreamed of following in her father's footsteps to become a neuroscientist. But at age nineteen, she attended a lecture by one of
Poland's most celebrated scientists, Andrzej Tarkowski, which
changed her life.[2] He 'was sitting in front of the room, no slides,
and he was just telling the story of how you manipulate embryos
... [and it was] magical'.[3] From then on, she made it her ambition
to understand how embryos develop. Few biological systems can
be as important and relevant to us as the genesis of human beings.

And scientifically, a special elegance of studying embryos is this: when looking at any other living tissue, it's hard to know about its history – the journey each cell has taken to reach its present condition, how its complexity has been arrived at – but when you study an embryo, you're starting from the very beginning.

After earning her PhD in Warsaw, studying embryos with Tarkowski, in 1995 Zernicka-Goetz moved to Cambridge University.[4] There she worked with Martin Evans, who was already renowned for discovering in 1981, with colleague Matthew Kaufman, a way to extract cells from a mouse embryo and grow them in a lab dish.[5] Logic alone tells you that an early embryo contains cells capable of becoming all different kinds of cells; otherwise there would be no way for an embryo to become a whole body. But what isn't obvious, and what Evans and Kaufman showed, is that these sorts of cells – embryonic stem cells – could be isolated, grown and manipulated in a lab dish.[6] This opened up the idea that embryonic stem cells might be used medically, to help replace or restore damaged tissue.[7] Aware of such potential, they had hurried to publish their results so that nobody would be able to gain commercial rights over the process.[8]

When Zernicka-Goetz arrived in Cambridge, she realised that, although it was vitally important to study how cells extracted from an embryo could become other cells in a lab dish, what was missing from these experiments was a sense of how cells move within an actual embryo, and how a cell's position in the embryo impacts its behaviour and fate. The basic question she wanted to address was this: does the position of a cell in an embryo determine what it will become, or does an embryonic cell adopt a speciality and then move to its correct location? To find out, she needed a way to watch the movement of cells in a living embryo, and to trace which cell derived from which other cell. As we've seen in Chapter One, a tool for doing this had just become available, in the form of the green fluorescent protein from jellyfish.

Zernicka-Goetz injected genetic material encoding the green jellyfish protein into one cell of a two-cell mouse embryo, so that this cell would glow green when illuminated under a microscope.[9] As the embryo developed, every cell derived from this first injected cell would gain a copy of the same genetic material and also glow green.[10] In later experiments, she also used chemical stains to see other embryo cells at the same time. By carefully tracing each cell's movement, and seeing which cell came from which other cell as an embryo develops, she discovered something she didn't expect and could hardly believe.[11]

In many organisms, including flies, worms and frogs, a fertilised egg becomes organised very quickly, so that when the fertilised egg divides, the two daughter cells are already different from one another. When these divide into four cells, each will again be slightly different, so that the individual cells already carry specific information towards what they will become. This opposed a long-held view that, for us and other mammals, an embryo was during the first few days an indiscriminate bolus of identical cells, and that only later did mammalian embryo cells begin to adopt a more specific identity. This standard view required that the early cells in an embryo be fully malleable and could become any other type of cell. In support of this, Tarkowski – Zernicka-Goetz's PhD mentor – had shown that if one cell in a two-cell mouse embryo was killed off, the remaining single cell could still lead to a healthy baby mouse. This implied that all the information needed to make a baby was still there in half the embryo.[12]

What shocked Zernicka-Goetz was that her experiments showed that the cells in a four-cell embryo were not, in fact, identical. Through a series of experiments, each beginning at about 6 a.m. and lasting about 20 hours, she found that each individual cell seemed to have already 'switched on' a genetic program that would shape its future character.[13] Two of the cells would give rise to all the cells of the mouse body, one cell would generate all the cells of the placenta (the organ where nutrients from the mother pass into the bloodstream of the baby), and the fourth cell would become the yolk sac (providing nourishment to an embryo until the placenta develops).[14] Nobody believed these results at first, and Zernicka-Goetz herself was sceptical to say the least: 'In fact, [these results] made me suffer quite a lot, because this model was against the dogma, it was against what I believed, what was proposed, against my PhD mentor ... and in fact, made me doubt myself.'[15]

By this time, she was leading her own research team in Cambridge, and they repeated the experiment in several ways. 'For years we couldn't believe that it happened ... [but] we imaged thousands and thousands of cells and followed them in this detailed way.'[16] To be clear, not everything is resolved even now: the development of a mammalian embryo is a complex progressive process, and it still remains contentious as to when, and to what extent, cells specialise.[17] Eventually, though, analysis of which genes are switched on and off helped support what she and her team saw under the microscope: that cells in a four-cell embryo are different from each other.[18]

It was in the midst of all this controversy, worry and self-doubt that a genetic test indicated a problem in her pregnancy.

It was her second pregnancy and it hadn't been planned. Though thrilled by the news, she took note of her doctor's advice to have a genetic test, because birth defects are well-established to be more likely in women over forty, and she was forty-two. So, two months into her pregnancy, she took a chorionic villus sampling (CVS) test, which involves taking a small sample of cells from the placenta with a syringe guided by an ultrasound scan. In the days after she received the results – indicating an extra copy of chromosome 2 in some cells – she racked her brain and scoured the scientific literature to try to understand what they meant for her and her baby.

In a CVS test, cells are taken from the placenta, not the baby directly, so Zernicka-Goetz reasoned that there were three ways to account for her test results. The best scenario, and the one she hoped to be true, was that the abnormality was limited to the placenta, arising at some time in that organ's development, and her baby was completely fine. But given that so many cells – about a quarter of those tested – were abnormal, it seemed unlikely that the problem would be confined to the placenta.

Another possibility, and the one she feared most, was that the problem had arisen in the baby, meaning that all or most of the baby's cells were abnormal. This could cause a miscarriage, or the baby could be born with any number of possible symptoms. The third possibility was that the situation in the placenta might accurately represent the situation in the baby: that *some* of the embryo's cells carried a problem. It dawned on her that nobody really knew what happens during the development of an embryo that contains some faulty cells – despite the fact that many women have to make very difficult decisions faced with precisely this scenario.

Talking to doctors working in an IVF clinic, she was shocked to learn that, in their experience, it was not uncommon for early human embryos to contain a mixture of normal and abnormal cells.[19] She directed her research team to study what happens to these so-called mosaic embryos. Scientific investigation and the development of her baby would progress alongside one another.

Because the deliberate generation of abnormalities in human embryos is not allowed, Zernicka-Goetz again turned to mouse embryos to perform her study. With her lab team, she examined what happened in healthy eight-cell embryos as compared to eight-cell embryos in which some cells were abnormal.[20] Once again, what she found was surprising, to say the least. By watching the development of these embryos under a microscope, she found that any abnormal cells that were positioned in the part of embryo that would normally go on to form the baby, died off as the embryo developed.[21] Healthy cells, meanwhile, would compensate for and replace the part of the embryo that had been lost. Some nearby healthy cells were even able to engulf the dead cells' remnants, seemingly deleting their existence entirely.

When these embryos containing some abnormal cells were implanted in foster mice, healthy babies were often born. Even if as many as half of the cells were faulty, embryos could correct themselves and babies were usually born healthy. If two-thirds of an embryo was abnormal, this would still lead to four in ten babies being born completely healthy. This meant that – at least for mice – there was enough flexibility in an embryo that the presence of *some* abnormal cells did not inevitably foretell problems with the baby's health.

Zernicka-Goetz's personal story unfolded faster than the science in her lab. A month after her initial test results, a second genetic analysis, this time from an amniocentesis, which samples a small amount of the liquid around the baby, had indicated that her baby was completely normal. And one January morning, her son Simon was born healthy. Her lab's results hadn't yet arrived. Even had they been available to her, the tests had been performed with mice and the abnormal embryos were created in an unnatural way, both of which would make it hard to translate the results into medical advice. Even now, some aspects of the process remain unclear, and decisions for women in this situation remain very difficult. In essence, Zernicka-Goetz was lucky that everything turned out well. But as this line of work continues, and especially if what happens to faulty cells in the development of a human embryo becomes more predictable, children may be born who might otherwise not have been.

Zernicka-Goetz says that she would not have begun to study what happens to an embryo with faulty cells had it not been for the worrying results of her own pregnancy tests. In 2019, I asked her son Simon, then aged twelve, what he thought about being the driving force for these important experiments. He said it was great, but he didn't seem too fussed about it. I think he had better things to be doing than talking to me about his mum's science.

These were Zernicka-Goetz's first scientific achievements, and arguably her greatest work was still to come. She and her team would soon carry out experiments extending the time an embryo can live in a lab, which forced us to confront the question of what – or rather, when – an individual is. And the broader picture is that our understanding of embryos, combined with a suite of other breakthrough technologies, CRISPR most obviously, allows us to influence who gets born to an unprecedented level. This is something that has been much discussed and debated already – the controversy is decades old – but what's different now is that many of the issues are no longer just abstract possibilities. In the last few years, what was once the realm of science fiction has become reality. And already some people are breaking the boundary of what's thought acceptable.

An egg cell is the largest human cell, but it's still only a little smaller than a full-stop on this page. After it's released from an ovary, it will die in about twenty-four hours unless it meets a sperm cell. But if it does meet a sperm, then everything kicks off. Within a day or so, the fertilised egg divides into two cells, then, a couple of days after that, four. Hormones then stimulate the lining of the womb to become receptive to the embryo for a brief time, known as the 'implantation window'. Six days after fertilisation, the minuscule embryo – consisting of about 250 cells and formally called a blastocyst – adheres to the lining of the womb and begins to burrow into the underlying tissue. If pregnancy fails, it's often at this point.[22]

Perhaps it shouldn't be surprising that this is a critical time: the choreography of events whereby an embryo successfully connects to its mother is astoundingly complex. As the embryo moves into the lining of the womb, cells from the embryo break down some of the mother's blood vessel walls. Blood leaks out into small pools to surround tree-like structures that emanate from the embryo. This is how the placenta begins, a temporary organ built to collect nutrients and oxygen for the developing baby, while removing waste. The baby's blood is never in direct contact with the mother's, but substances pass back and forth across thin membranes that separate the two. Construction of the placenta is very special: pretty much the only other human cells known to be able to break down blood vessel walls and reconstruct blood flow are cancer cells.

At the time the placenta begins to form, the foetus itself is a hollow ball of cells as big as a poppy seed; structured but far from being anything body-like. Around fifteen days after fertilisation, a clear top, bottom, front, back, left and right develop. After eighteen days, two small tubes appear. The two tubes merge a few days later and by twenty-two days, as if by magic, the single tube begins to beat.[23] This is the baby's first organ: its heart, needed to pump nutrients throughout its developing body.

Of course, all of this is hidden from view. The closest we come to knowing when somebody has become pregnant is with a test that can detect the presence of hormones about eight days after fertilisation, and more reliably a few days later than that. A three-week-old foetus can be picked up with an ultrasound scan, but an embryo's first days are extremely difficult to detect, let alone study in detail. Historically, knowledge of such early development was acquired by studying animals. The first microscopic view of a developing heart, for example, was provided in the late 1600s by the Italian biologist Marcello Malpighi examining chick embryos.[24] More recently, the anatomy of human embryos has been described using collections of foetal tissue obtained from surgery and abortions, the largest of which is the Kyoto Collection, holding some 45,000 specimens, the majority having been obtained between 1962 and 1974.[25] More recently still, detailed knowledge of what happens in the first few days of our lives has come from studying embryos donated from women who no longer need them following IVF.

IVF is unquestionably one of humankind's most revolutionary scientific achievements, not only helping treat infertility but also beginning our complicated journey through the myriad new possibilities of selecting, and even editing, our children's genetic inheritance. In 1959, the Chinese-American scientist Min Chueh Chang was the first to achieve IVF in a mammal. He transferred a black rabbit's fertilised egg into a white rabbit, leading to the white rabbit giving birth to a black litter. The first human pregnancy with IVF was reported in Australia in 1973, but led to a miscarriage.[26] A successful pregnancy from IVF happened five years later, in the UK.

This profoundly important achievement was largely down to five people: Robert Edwards, Patrick Steptoe, Jean Purdy and

Lesley and John Brown. Edwards and Steptoe met at a scientific meeting in London in 1968, and bonded over their determination to solve infertility. Edwards – 'the competitive second son of a working-class family', as he described himself [27] – was one of the world's leading scientists studying reproduction at the University of Cambridge. Steptoe was the Director of the Centre for Human Reproduction in Oldham, in the north of England, having failed to obtain the consultant's post he wanted in London.[28] Purdy helped run Edwards' lab and became, in essence, the world's first IVF nurse. Sadly, she died from cancer in 1985 at the age of 35. Purdy's role in developing IVF has often been overlooked, despite Edwards always championing her.[29] Lesley and John Brown were the couple from Bristol who wanted to have a baby but seemingly couldn't. They had tried for ten years and Lesley had become depressed.[30] Of course, a sixth person was also crucial: Louise Brown herself, the world's first 'test-tube' baby, born on 25 July 1978.[31]

Steptoe had predicted that Brown's birth would prove more important than mankind landing on the Moon – and arguably, he was right. In the build-up to it, however, the idea that it should even be attempted was hugely controversial. James Watson, the Nobel laureate who co-discovered DNA's double-helix shape, thought that using IVF for pregnancy was far too great a risk: 'All hell will break loose, politically and morally, all over the world,' he told a US Congressional subcommittee.[32] Max Perutz, another Nobel laureate and one of Cambridge's most renowned scientists for his work on haemoglobin, said to the press that if an abnormal baby was born, the guilt would be tremendous, and 'the idea that this might happen on a larger scale – a new thalidomide catastrophe – is horrifying'.[33] Edwards and Steptoe were affected especially directly by the UK government's funding body, the Medical Research Council, not supporting their initial proposal to use IVF for pregnancy. The Council saw infertility as relatively unimportant and, astonishingly from today's perspective, when scientists are usually encouraged to talk about their work in public, Edwards and Steptoe were also accused of seeking too much media attention for the topic. One scientist the Council consulted was of the view that

> Dr Edwards feels the need to publicise his work on radio and televi-
> sion, and in the press, so that he can change public attitudes. I do not
> feel that an ill-informed general public is capable of evaluating the
> work and seeing it in its proper perspective. This publicity has antag-
> onised a large number of Dr Edwards' scientific colleagues, of whom
> I am one.[34]

Evidently, Edwards and Steptoe were pioneers in the public discussion of science as well as in developing a technology that has by now has led to over eight million babies being born. Taking into account that many people born by IVF will later become parents themselves, it's been estimated that by the year 2100, around 1–3 per cent of all humanity will owe their lives to reproductive technology.[35] For such an enormous accomplishment, it's surprising and disheartening that it took thirty-two years for a Nobel Prize to be awarded for the development of IVF, by which time Steptoe and Purdy had both died and its sole recipient, Edwards, was eighty-five years old and too frail to attend the ceremony.

For pregnancy using IVF, embryos are transferred to the womb after being cultured in a lab dish for two to six days. Unused embryos can be frozen for future attempts at pregnancy, or used for research, with the parents' consent. For research, they can be cultured longer, but until recently they could not be kept alive for much more than about a week. That is until Zernicka-Goetz found a way to culture embryos far longer than anybody had done before.[36] This breakthrough was transformative scientifically, as we will see, but also reignited different people's passions and feelings about the legal limits on human embryo research.

The fourteen-day limit on how long a human embryo can be kept alive in a lab dish was first suggested in 1979 by a US ethics advisory board, and then endorsed in 1984 by a report for the UK government, known as the Warnock Report, named after the committee's chair, the moral philosopher Mary Warnock.[37] A few countries, including the UK, Spain and Australia, have since made it a criminal offence to grow a human embryo any longer.[38] The Warnock committee had spent two years grappling with multiple conflicting interests over a technology that at the time was far from established.[39] 'Disputes were on the whole civilised,' Warnock recalls in her memoir, but 'I think that by the summer of 1984, tempers were a bit frayed.'[40]

The committee's success was, at least in part, down to answering a question subtly different from the one that is usually asked. The central problem – as the press would always have it – lay in finding an answer to the question, 'When does life begin?' Warnock's committee took the view that this was not a question of fact, as it seems at a glance, after all, but something which had to be decided. And since a live human embryo in the lab was something that had never existed before, they reasoned that really, the crucial new question was this: how should we regard this new entity, a living human embryo outside the uterus? In other words, their focus was on deciding when a human embryo in a lab dish reaches the point at which it needs protection.

Not everyone's opinion could be reconciled, but the committee found consensus in setting a limit to the length of time any human embryo can be allowed to live in a lab.[41] Their decision – the fourteen-day rule – was justified in several ways. In a fourteen-day-old human embryo, there's no sign of a nervous system, which would be a prerequisite for feeling or thought. Also, many embryos are naturally lost during the first two weeks. On day 15, moreover, a groove appears in the disc-shaped embryo, called the primitive streak. This coincides with the embryo no longer being able to split and develop into twins. Arguably, before this moment, an embryo can't be considered an individual, because if it was, how could it still be able to split and become two individuals? From this logic, the presence of the primitive streak, on day 15, can be taken as the moment at which a unique human being has come into existence.

One argument against a fourteen-day limit is that an embryo cannot possibly experience pain until much later in its existence. Neurons that transmit signals from the spinal cord to the part of the brain where pain can be perceived do not develop until a foetus is around twenty-three to twenty-four weeks old. And an argument against the idea that a human embryo becomes a distinct individual on day 15 is that embryos used for research are never destined to become a person anyway.

On the other hand, as decreed by Pope Pius IX in 1869, the Roman Catholic Church considers a person to have been created at the moment an egg is fertilised by a sperm. Interestingly, this position of the Catholic Church was directly influenced by scientific technology.[42] In the early seventeenth century, microscopes could just about pick out the outline of sperm. Some scientists at the time theorised that tiny human beings must exist inside the sperm's head. This view – wildly wrong of course – implied that men could take credit for creating the next generation, while women served merely to nourish and enlarge a person's body.[43] Catholic theologians took the theory of a preformed human body inside sperm to imply that personhood must begin at conception.

Roughly speaking, Hinduism also holds that life begins at conception, but allows for an embryo's destruction in some situations. Judaism considers an embryo's status to increase over time, and says a soul may enter on its fortieth day. Many Islamic scholars agree with this view, although the Muslim World League considers ensoulment to happen later, 120 days after fertilisation. From this – a very small snapshot of religious views – it is blatantly difficult to take on board the world's diverse values in formulating global rules for embryo science. Anyone wishing to extend the current fourteen-day limit might hold back simply because widespread public discussion would ignite all sorts of strongly held feelings, which could lead to any possible outcome, including the limit being reduced rather than extended.[44]

At the time Warnock's recommendation was adopted, no science was actually restricted by a fourteen-day limit, because it was technically impossible to preserve an embryo outside of the womb for this long. So the restraint primarily served to maintain the perception that science was being morally controlled. However, breakthroughs by Zernicka-Goetz and others, published in 2016, have reignited the debate.[45]

Zernicka-Goetz was driven by the fact that what happens to an embryo after its first week had been exceptionally hard to study: 'I wanted to take a look inside this "black box", [to] see what was going on.'[46] Her team began with mouse embryos. Day after day, her team tested countless conditions of hormones, nutrients and growth factors that might keep the embryos alive longer than anybody had achieved before. The days became months. Not only did they change the broth the cells were kept in but they also tested, for example, whether the embryos might survive better if placed on a soft gel rather than the usual hard plastic dish – it turned out not to matter. Eventually, they saw a mouse embryo live in a lab dish longer than anybody had seen before, a couple of days longer than the time it normally took for embryos to be implanted. But this success was short-lived, because the method proved unreliable; it seemed to work one time and not the next. The iterative process of tweaking every-thing continued for many months more, until the procedure was robust.[47] Finally, it worked. Then the next step was obvious – to test their method for mouse embryos on ones that were human.

One day in May 2013, they began to culture two human embryos donated from an IVF clinic.[48] Amazingly, one of them started to develop. As this human embryo continued to live past eight days, it dawned on Zernicka-Goetz and her team that, because nobody had ever seen a living human embryo in a lab dish past this point, they had no way of knowing if what they were about to witness would be anything like what happens in the womb. By day 11, however, the embryo began to self-organise, and looked similar to what was shown in textbooks based on earlier studies of samples collected from operations.

On day 12 they terminated the project, and in every future experiment they never went beyond day 13, because of the international agreement born out of the Warnock recommendation, enforced by law in the UK. Around the same time, and in collaboration with Zernicka-Goetz, a team in New York, led by the Iranian-born scientist Ali Brivanlou, achieved a similar feat.[49] Brivanlou had sent one of his team to Zernicka-Goetz's lab to learn their method for keeping mouse embryos alive and then, back in his lab, tweaked the method for human embryos.[50] Brivanlou vividly remembers the moment he met with his team to decide whether or not they should kill the embryos as the fourteen-day deadline approached. In the USA, the cut-off date is a guideline rather than a law, so continuing wouldn't have been illegal, but Brivanlou decided to terminate the experiment. Without naming names, he told me there were tears in the team.[51]

These two lab team's achievements were voted by readers of *Science* magazine as 2016's breakthrough of the year, because their work opened up a new way of studying the earliest phase of human development, the beginning of human life. The feat itself was important – 'mind-blowing', Brivanlou says[52] – because the discovery that an embryo can survive in lab conditions for so long, seemingly 'implanting' itself against the bottom of a lab culture dish, was unexpected. The implication is that an embryo is self-sufficient for some time after it implants, requiring little, if anything, from the mother's tissue at first.

By thirteen days, however, there were signs, at least in Zernicka-Goetz's lab, that the embryos needed something other than the culture they were in. Perhaps by including maternal tissue or complex human-made materials, they could be made to survive longer. It is highly unlikely that the fictional hatchery in Aldous Huxley's *Brave New World*, used for growing cloned humans in incubators, will ever be possible, but as for how long a human embryo could *possibly* live outside a womb, nobody knows.

Brivanlou, for one, would like to try to grow human embryos for longer, up to twenty-one days.[53] There's so much that can be learnt from watching embryos develop, he says: from understanding the appearance and disappearance of countless structures as a new human begins, to figuring out what is happening when human development goes wrong. To circumvent restrictions, he and others are also studying so-called synthetic or artificial embryos. Essentially, these are clumps of stem cells treated so that they develop basic structures of actual embryos, without there ever being the slightest chance of their becoming a body. At least for the moment, artificial embryos do not pose any major ethical issues. But as for growing real human embryos, Brivanlou knows it wouldn't be right to push ahead unilaterally. Human embryo research is morally, culturally and politically controversial, and society spans every conceivable opinion. There is a consensus in place, but it's fragile.

Still, deciding how long a human embryo should be cultured for is not even the most pressing or challenging issue we now face. Recent advances in IVF have thrust to the fore other, even more complicated dilemmas.

Making a baby without sex is a vastly more sophisticated process today than it was when Louise Brown was born in 1978. Our understanding of the relevant science has advanced dramatically, and now there are a host of opportunities to make interventions and decisions, raising many difficult issues for parents and society.

The IVF process begins with daily injections. For about two weeks, a woman injects herself with hormones so that her eggs mature. The injections amount to a hormone dose higher than would naturally occur, causing several of her eggs to mature at once. Using a needle, passed through the vagina and guided by ultrasound, her eggs are retrieved. One by one, each is gently drawn out using light suction, until a dozen or so are collected over about twenty minutes, while the woman is sedated with anaesthetics. The eggs are usually surrounded by other small cells called cumulus cells. In a nearby lab, the collected eggs are examined under a microscope and graded – essentially for their looks – taking into account whether or not a good number of cumulus cells are present and whether or not the sample has a healthy-looking texture. Fresh semen is usually collected on the same day, at home or in the clinic.

Before being allowed anywhere near an egg, the sperm are often washed. This was first done in the mid-1990s when it was discovered that HIV could feasibly be passed on through the father's semen to the mother or child. Nowadays, the process is used not only to remove infectious agents, but also because some components of seminal fluid can inhibit fertilisation when carried out *in vitro*. On the face of it, washing minuscule sperm might not sound easy, but there are several ways to do it. Commonly, semen is diluted in a solution containing antibiotics and protein supplements before being spun in a centrifuge – a device something like a washing machine but able to whizz around much faster – so that the sperm concentrate at the bottom of the tube. The liquid is siphoned off and the sperm are resuspended in a fresh solution – *voilà*, washed.

One IVF clinic in California offers a menu of sperm washes. They run from basic to premium. The process just described is basic. For a premium wash – more expensive, of course – sperm are centrifuged in a test tube containing layers of liquid that create a density gradient. This helps purify healthy sperm because dead sperm gather at the top of the tube and can be discarded. Another option – price on application – involves leaving semen in a tube full of culture broth. An hour or so later, the top part of the liquid, containing sperm capable of swimming up the tube of their own accord, is extracted, leaving dead or non-moving sperm at the bottom.

With the sperm washed, fertilisation is attempted in one of two ways. Thousands of sperm can be mixed with an egg cell in a lab dish and left in an incubator for a few hours, in the hope that, by chance, fertilisation will happen. Alternatively, a needle can be used under a microscope to insert a single sperm directly into an egg cell – a procedure called intracytoplasmic sperm injection – relieving the sperm of the task of finding and entering the egg of its own accord.

The next step is to give the fertilised egg time to grow. Again, there are countless options: the best way to culture a fertilised human egg for successful pregnancy is the topic of well over a thousand scientific papers.[54] A small industry has grown up marketing culture broths as optimal for human embryo growth, each with varying amounts of glucose, amino acids, vitamins, antibiotics or growth factors.[55] There are other variables too: levels of carbon dioxide and oxygen, temperature and humidity can all be adjusted in the incubator where the fertilised egg is kept. Movement might also be beneficial, so sometimes the developing embryo is kept on a gently rocking platform.[56] All of this almost certainly affects an embryo's growth and its potential for pregnancy, but nobody knows what's really optimal, and each clinic has its own set-up.

To grade embryos for their likely chances of successfully leading to pregnancy, an embryologist looks at them under a microscope. They look for the cells to appear smooth and round, for example, and to see if all of the cells are dividing. A bulge, or 'bleb', as it's called in scientific texts, can protrude from one or more of the embryo's cells, for reasons that aren't clear, and if this is happening a lot, the embryo gets a low grade. If the embryo grows to a couple of hundred cells or so, an embryologist can also assess whether or not it has gained the right structure of a hollow ball. These judgements are an art as much as a science. Embryologists make the best decisions they can, but it's hard to pick out which are really most likely to lead to successful pregnancy just by looking at them.

In 2019, the ability of embryologists to assess embryo quality was compared with that of artificial intelligence (AI).[57] The test was whether an individual embryologist's assessment of the quality of an embryo matched that of the majority of embryologists more or less often than the AI. The software, based on an image-recognition system developed by Google, was fed 12,000 pictures of embryos already categorised as poor or good, to find patterns separating the two groups which it could look for in other embryos. By analysing the images in all sorts of ways, the software learnt to pick out subtle and complex differences in shapes and textures, which would be hard or impossible for an embryologist to know how to assess.

So the outcome of this mini Kasparov-versus-Deep Blue duel was that AI won, at least in the sense that AI was more consistent.

Individual embryologists varied a lot in their scores, but the software was virtually always in agreement with the majority decision. Of course, this doesn't prove that AI could help maximise a woman's chances for pregnancy – not least because the majority human decision might not have always been right, and this wasn't set up as an actual clinical trial. But it suggests AI could help. In a future upgrade, the software might be able to categorise embryos more precisely, picking out those with specific chromosomal abnormalities, for example.[58]

To more accurately assess the health of an embryo, a biopsy can be taken. Unlike taking a sample of bone, liver, kidney or other tissue from an adult body, a biopsy from an embryo doesn't require a surgeon but an embryologist who can work with the most fragile of live samples, using pipettes and a minuscule needle under a microscope. From a biopsy of an embryo, its genes can be scrutinised – a process called pre-implantation genetic diagnosis or PGD.

To take a biopsy from an embryo, the first step is to pierce the thick, transparent membrane which surrounds it. This can be done in several ways, with a needle, with pulses of laser light or with a small squirt of concentrated acid, and each has its pros and cons. A laser is easy to use, for example, but also heats up the liquid around the embryo, which might be a concern even though there is data to say it's safe.[59] Whichever method is used, the embryo has to be pierced just right: too small a hole and a cell can't easily be pulled out; too large and cells may come out of their own accord and be lost. A broth lacking calcium and magnesium ions is sometimes added to reduce how tightly the embryo cells are stuck to each other. Then, with the embryo held steady under a microscope, a pipette can gently suck out one or a few cells. Yet again, there's an alternative: a pipette can be used to push against the embryo's outer membrane, the pressure causing a cell to be expelled. Either way, there's a chance that the cells being taken, or the embryo itself, is damaged in the process and has to be discarded. But being overly careful isn't a good idea either, because the speed of obtaining the biopsy is crucial too; living embryos shouldn't be out of their incubators for long. Then, while the biopsy is being analysed, the embryos are frozen, each potential life suspended, while science decides which of them might be born.

Apart from those who oppose any level of intervention in human reproduction, few would argue against allowing parents the opportunity to screen embryos following IVF in order to avoid a single genetic variation that would otherwise certainly lead to progressive motor and mental difficulties, as is the case with Huntington's disease, for example. But for each prospective parent, none of the choices involved are easy. Decisions require taking a position on the moral status of an embryo and deciding what to do with embryos that aren't going to be used – they can be destroyed, frozen or used in research. The cost of PGD is also a problem; most US health insurance companies will not pay for it.

Things become even more complicated when considering screening embryos for a genetic variation that won't inevitably cause a problem. Today, there are over 400 conditions which can be tested for in the UK.[60] Many of these are genetic variations which carry *some* level of risk, the precise level of which is often not clear.[61] Certain genetic variations only cause problems late in life, by which time other treatments could feasibly be available. As well as this, the effects of most genetic variations are complex. A gene variant which correlates with an increased risk of a particular autoimmune disease, for example, also correlates with being better able to fight off HIV.[62] Needless to say, there is no such thing as an ideal genetic inheritance; human diversity is fundamentally important. The problem of using PGD to select embryos for implantation is that it forces us to answer one of the most vital and fraught issues of our time: what really is a genetic disorder?

Paula Garfield worries about the idea that deafness is seen as something which might be genetically screened against. Garfield and Tomato Lichy, who are both deaf themselves, were thrilled when their first child was born, who also happened to be deaf. As Andrew Solomon writes in his book, *Far from the Tree*, about children and identity, 'the general culture feels that deaf children are primarily children who *lack* something: they *lack* hearing. The Deaf culture feels that they *have* something: they *have* membership in a beautiful culture.'[63] It's very important, Garfield says, that society embraces human diversity, and who wouldn't agree? As she told me in 2019:

> It is Doctors and Audiologists who are the ones to break the news to hearing parents that their baby/child is deaf. They usually start with – 'I'm very sorry to say that your baby/child is deaf.' As soon as hearing parents hear the 'I'm very sorry' part then they immediately think that the deafness is something negative, that this is bad news. For me, the message should be more neutral – 'Your baby/child is deaf but don't worry, there's lots of support and services available out there for you to access.'[64]

Garfield and Lichy hit the headlines across several newspapers in 2008, for arguing that we need to be careful in what we screen embryos for following IVF: 'We're proud, not of the medical aspect of deafness, but of the language we use and the community we live in,' Lichy said to the press.[65] They did not, as some headlines seemed to imply at the time, wish to have a deaf baby deliberately. Rather, they wanted to stress that if they had to go through IVF to have a child, they would not want an embryo to be discarded *just because* the baby would be deaf. For trying to make this point, Garfield says, 'We got a huge backlash on social media [and] it adversely affected my mental health and well-being as well as my family and relationships.'[66]

Many of us have prejudices we must work to suppress. Hearing loss caused by an infection, for example, is something we all agree should be prevented. But in terms of selecting embryos following IVF, as Julian Savulescu, a philosopher based at the University of Oxford, has argued, being born deaf is not harmful:

> Is that child worse off than it would otherwise have been (that is, if they had selected a different embryo)? No – another (different) child would have existed. The deaf child is harmed by being selected to

exist only if his or her life is so bad it is not worth living. Deafness is not that bad.[67]

As Garfield puts it: 'The initial focus was being able to screen for illnesses which are life-threatening or that could cause early deaths. Deafness is not a life-threatening condition, you can't die from being deaf.'[68]

In 2002, the lesbian deaf couple Sharon Duchesneau and Candace McCullough chose a sperm donor with five generations of deafness in his family.[69] In this way, the couple did, in effect, choose to have a deaf baby deliberately. That science helped this baby be born is what makes this controversial. In everyday life, anyone is free to choose a sexual partner with a view to what life might be like for any prospective children they might have together. But the issue of which genes we should be allowed to select for or against is entwined with the science of what's possible, when and how. No simple rule works: we cannot say, for example, that any selection of embryos is unacceptable, because in other circumstances we have long allowed selection, and much later during the developmental process. Screening for Down syndrome during pregnancy is common, for example, and we allow parents the freedom to choose whether or not to have a child with Down syndrome.

I do not have the answers, and my own opinion is no more important than yours. The crucial point is that new science is opening up an unprecedented number of options for how we conceive and for the fate of our children. Our actions must fall within each country's laws and regulations, but even so, it's possible to travel to places with different rules (or with a less strict application of the rules).

We are responsible for our children in all sorts of ways, influencing the food they eat, the school they attend, the hobbies they take up and the friends they keep. Selecting their basic genetic make-up, however, is a whole other level of influence. The decisions we make will not change the nature of our species – this is not about engineering humanity in any global sense. But it is about new science leading us to make important decisions in our own lives and for our children's lives. Where once things were left to chance, progress has brought us choice.

In 2018, a controversial breakthrough in embryonic manipulation caused international uproar, and made clear how momentous those choices have become.

In November 2018, the Chinese scientist He Jiankui, then based at the Southern University of Science and Technology, Shenzhen, claimed to have used CRISPR – a technology for gene-editing based on DNA sequences in bacteria and archaea – to edit the genes of non-identical twin babies.

It had already been rumoured that He had obtained local permission to try to produce genome-edited human embryos, which seemed feasible given that he had previously presented data on editing mouse and monkey embryos.[70] For this reason, he had been invited to give a talk at the Second International Summit on Human Genome Editing at the University of Hong Kong. Late in the afternoon on 25 November, two days before the meeting was due to begin, organisers were alerted that He had emailed Jennifer Doudna, one of the pioneers of CRISPR technology,[71] to inform her that two genome-edited babies had been born.

Doudna had been studying CRISPR – shorthand for the Clustered, Regularly Interspersed, Palindromic Repeats observed in the gene sequences of bacteria – since the time when almost nobody had heard of it. While she was growing up, her father, a professor of literature at the University of Hawaii, brought lots of books home. One of these, James Watson's *The Double Helix*, inspired her to become a scientist, because it showed her the human endeavour behind the textbook description of facts.[72] After studying chemistry, she set up her own lab at Yale in 1994, and in 2002 moved to Berkeley, California. It was there that she first came across CRISPR. In 2006, Jillian Banfield, another professor at Berkeley, phoned her out of the blue. She had found Doudna through a Google search for someone local with the right experience to help her study CRISPR. Doudna soon found herself enthralled by the topic. The reason they were both excited was that in 2005, three publications had reported that the CRISPR part of the genome in bacteria contained sequences which matched viruses known to attack bacteria.[73] This hinted that CRISPR could be involved in the immune defence of bacteria, helping them fight off viruses. For their role in understanding CRISPR, Doudna and the French scientist Emmanuelle Charpentier – said to be 'so resourceful she could start a lab on a desert island'[74] – won a Nobel prize in 2020.

Of course, this snapshot belittles the whole CRISPR story, which unravelled over twenty years across labs in at least nine different countries and could easily fill a book in itself.[75] The story amounts to another example of how a discovery of great medical significance began with relatively obscure research, in this case seeking to understand strange genetic sequences in microbes. The picture we have now is that microbes do indeed use the CRISPR system to attack invading viruses. This works by tagging viral genes for destruction by enzymes which naturally occur in bacteria. This is fascinating in its own right, but the reason this basic science is considered an important medical breakthrough is that the CRISPR system can also be repurposed to edit the genome of animal cells, including human cells, in any way we choose. In effect, CRISPR provides a way for almost any biology lab to switch off or edit genes inside cells. This includes a way to manipulate – not just screen – the genetic make-up of embryos.

In 2015, the use of CRISPR to edit genes in human embryos was first reported.[76] In that study, the Chinese scientist Junjiu Huang tested CRISPR in human embryos which were anyway defective, because they contained an extra set of chromosomes, ensuring that no babies could have resulted from this work. As it turned out, the experiment was not as successful as had been hoped, because mutations occurred in genes other than the one he had intended to edit. Reaction to the work among the scientific community was mixed, but one thing seemed clear: if CRISPR was ever going to be used as a means of gene-editing embryos, its accuracy had to be improved at the very least.

Many scientists have called for a moratorium in using CRISPR for manipulating the genome of human embryos. Indeed, for a while there was a worldwide moratorium on gene-editing human embryos, but little is in place – and arguably, little can ever be in place – to really stop it. In November 2018, it suddenly became apparent that He Jiankui had decided to override the scientific consensus.

In the days leading up to He's appearance at the summit at the University of Hong Kong, a journalist for the *MIT Technology Review* had unearthed some information about He's work from a Chinese clinical trial registry.[77] On 25 November, the *MIT Technology Review* published what they knew so far.[78] That evening, five well-produced videos made by He appeared online. In them, he described his work and confirmed that gene-edited twins – dubbed 'Lulu' and 'Nana' – had been born.[79]

Nervous that the press would find him, He left the hotel where all the other conference speakers were staying and moved to a secret location. Hong Kong University was later told his whereabouts, collected him and brought him to a hidden room, where he waited until being called into the auditorium, to face conference delegates and around 160 journalists. At the podium, He immediately apologised that his results had been leaked rather than first presented to a scientific audience. He also thanked his university, but said that they were unaware of what he was about to say.

For about twenty minutes, He went through his prepared talk. He spoke about some of his preliminary research with mice and monkey embryos before describing what he had done with human embryos. In human embryos, he had chosen to disable a gene called *CCR5*. This gene is naturally disabled in about 1 per cent of Northern Europeans, and protects them against infection by many strains of HIV (because most strains of the virus use the protein encoded by *CCR5* to gain entry into human cells). This, He said, was the rationale behind his work – to make children resistant to HIV.

The gene-edited babies' father was HIV-positive and the mother wasn't. Although having an HIV-positive father does not strongly impact the baby's chance of being born with HIV, He hoped that these two babies would be resistant to HIV throughout their lives. He said this was especially important in China, where being HIV-positive carries a considerable stigma. He had attempted to ensure there were no off-target genetic effects, and clarified that two embryos were edited independently and implanted, leading to non-identical twins.

After his prepared speech, He was joined onstage by two scientists, Robin Lovell-Badge from the Crick Institute in London and Matthew Porteus from Stanford, who asked questions – collegiately but carefully. From this inquisition, we learnt that eight couples had enrolled in He's clinical programme, one had dropped out, and that he had worked on thirty-one embryos in total. David Baltimore came to the podium next. With the gravitas of being an eighty-year-old Nobel laureate, he said that the work had not been done with any level of transparency – 'We only found out about it after it's happened' – and that there were more pressing medical needs than providing one person with some protection against HIV infection. Standing a few metres from where He was sitting, Baltimore declared the work irresponsible.

He's claim has yet to be formally verified, but it is certainly feasible that he told the truth about what he has done.[80] His work quickly received scathing criticism from all sorts of scientific, academic and medical institutions and numerous governmental bodies – because he acted outside everyone's ethical guidelines for human embryo experimentation and because the procedure is dangerous. As well as this, the actual mutations He made in the two babies are not exactly what he had desired. He had sought to make a specific mutation in the *CCR5* gene that is known to occur naturally, but in fact – and it's not clear why – the two implanted embryos ended up with slightly different mutations in the *CCR5* gene. The effect of these mutations is not certain: they may well confer some level of HIV resistance, but may also have other consequences in the immune system, where this gene is known to be important, or elsewhere in the body. There's some evidence that *CCR5* could have a role in the human brain, for example.[81]

Still, He can't be dismissed as a madman or 'cowboy scientist' in the way some early versions of the story suggested, and not every scientist has condemned his work. Trained in the USA, He had returned to China in 2012 through a prestigious programme run by the Chinese government. Stephen Quake, He's postdoctoral adviser at Stanford, told the *New York Times* that he was 'bright and ambitious'.[82] George Church at Harvard, who also played a role in developing CRISPR technology, said, 'As long as these are normal, healthy kids it's going to be fine for the field and the family.'[83] Many scientists have been particularly critical of the lack in transparency in He's work, but he may have had to work with some level of secrecy to get around Chinese rules that forbid anyone with HIV using IVF to have a baby.[84]

Robert Edwards – the IVF pioneer and Nobel laureate – is He's hero.[85] Edwards, with Patrick Steptoe and Jean Purdy, went ahead with a new medical procedure in the face of criticism, and using IVF for pregnancy was a huge risk at the time. But it worked and turned out to be truly revolutionary. Edwards and his colleagues did try to act ethically, but procedures and guidelines weren't as well defined as they are now.[86] Louise Brown's mother hadn't realised just how pioneering IVF was until she was about six months pregnant and saw herself discussed in the newspapers.[87] Whether or not IVF itself was tested too quickly is arguable. Perhaps the only answer is that it worked. But genetically editing human embryos for pregnancy is widely regarded as a step too far, or at least too soon.

Following the Hong Kong meeting, He disappeared from public view. His research programme was halted, and there is some evidence that, at least at one time, he was held in a guarded apartment on the university premises.[88] Over 120 scientists, including many Chinese, circulated an open letter urging immediate legal action and global discussion now that this particular Pandora's box had been opened. In March 2019, the World Health Organization? set up a committee to debate the issues. For some time, it was unclear whether or not the Chinese government would formally press criminal charges against He, but eventually it became clear that it did. In December 2019, a court fined He three million yuan ($430,000), and then in January 2020 he was sentenced to three years in prison.[89] Two of He's colleagues were given lesser fines and sentences.

Several national academies for medicine continue to debate the issues. Rules and guidelines will be announced, but genetic editing of human embryos is much more difficult to control than, say, the proliferation of nuclear weapons – because gene-editing cannot be monitored, guarded or stopped in any easy way. We will be grappling with the use of this technology for at least a century to come.

What emerges from all this is that we are about to witness a huge sweeping change in what is possible for the way we have children. Rules governing the use of this new science are almost certain to vary between countries and cultures, and be hard to enforce. We are each going to have to make our own decisions – with our lives and our children's lives affected deeply by what we do. Things are moving so fast that it's impossible to know, in a hundred years' time, say, how many children will be born because of the new science emerging now. And this is still only the beginning.

3 A Force for Healing

Marco Polo describes a bridge, stone by stone. 'But which is the stone that supports the bridge?' Kublai Khan asks. 'The bridge is not supported by one stone or another,' Marco answers, 'but by the line of the arch that they form.' Kublai Khan remains silent, reflecting. Then he adds: 'Why do you speak to me of the stones? It is only the arch that matters to me.' Polo answers: 'Without stones there is no arch.'

Italo Calvino, *Invisible Cities*

Leonard 'Len' and Leonore 'Lee' Herzenberg married in the summer of 1953, when he was twenty-one and she was eighteen. 'Our parents thought we were too young, too innocent, too poor, and too crazy,' Lee recalls.[1] But for over fifty years, until Len died in 2013, they led ground-breaking research together.

At the time they began their scientific journey, it was common for labs to build their own instruments rather than buy everything they needed. This was an era when the development of tools and techniques was widely recognised as a vital first step to a scientific breakthrough. Len and Lee's major achievement was developing a scientific instrument, used today by almost every biology lab and every hospital, to count, sort and analyse the body's cells.

In 1959, Len was recruited to Stanford to set up a new lab and Lee, who was at an earlier stage in her career, went with him. She had planned to look for a job in a different department or finish her studies. But soon after they arrived, it became clear that Len needed help, so they began working together. Initially,

Len directed the research, but in time they both became known for their brilliance separately, as well as together.

They loved each other and they loved science. DNA's double-helix shape had recently been discovered, and genetics was especially fascinating to them. But also, they thrived on 'getting lots and lots of things done'.[2] In Stanford, they also became involved in studying the immune system, just as its complexity was opening up. Many different types of immune cell had been discovered – some that were especially good at engulfing bacteria, while others could kill virus-infected cells. But our understanding at the time was messy; it wasn't clear what some types of immune cells did, while others had yet to be discovered. One of the tasks Len faced was to count how many of each kind of cell there was in a sample that contained numerous varieties. To differentiate them from one another, he used a technique which made each type of cell take on a particular colour, effectively 'labelling' them and thus allowing him to count them by sight. But counting cells one by one under a microscope was very laborious and, as Lee recalls, 'Len has very bad eyes … [and] he hated microscopy.'[3] Len realised that it would be much easier if there was some kind of machine which could count labelled cells for him.

Len was a biologist, not an engineer, but his upbringing – in Brooklyn, the son of second-generation immigrants – gave him the boldness and determination he needed. During Len's child-hood, his father worked in a clothing store, his mother was a legal secretary, and for a while they sent Len to a boarding school in upstate New York, which he never really liked. For a long period, he travelled by train to visit an orthodontist each week in Brooklyn. By navigating New York City's public trans-port system alone at the age of ten and eleven, Len gained the confidence that he could always do whatever's necessary.[4]

The most important thing behind Len and Lee's success, however, was that they were each other's champion and friendly critic. They met when Lee began studying at Brooklyn College, and Len was in his final year there. They perhaps bonded because they had both grown up in New York to Jewish families with ancestry in Eastern Europe and Russia.[5] As Len left for Caltech, they agreed to marry in three years' time, when Len's graduate studies and Lee's undergraduate studies would both be finished.[6] But that plan didn't work because they were too lonely when apart. Neither Skype nor email had been invented yet, and they could only keep in touch with expensive long-distance telephone calls. So Lee quit Brooklyn College and they got married. They loaded their stuff into a car that Len's parents gave them, and set off together on a 3,000-mile road trip across the USA, from Brooklyn to Caltech, for love and science.

Amazingly from today's perspective, women weren't formally admitted to Caltech on any undergraduate or graduate programmes at that time. But thankfully, the faculty itself thought women were worth educating and Lee was allowed to sit in on whatever courses she wanted. Each professor gave her a letter certifying that she had taken the course, and graded her performance as they did all the other students'. Lee now has the exceptionally rare distinction, perhaps unique, of having become a professor at Stanford University without ever having formally graduated from college.

Len and Lee's journey to Stanford from Caltech took them via Paris and the National Institutes of Health in Bethesda.[7] In Paris, Len worked in Jacques Monod's laboratory at the Pasteur Institute, where 'every day was an intellectual feast.'[8] During this time, Lee helped Len in the lab while she, in her words, learnt 'to balance being a mom with being a scientist'.[9] She often brought their newborn baby into the lab during the afternoons. Her recollection is that women were much more welcome in labs in France than in the US at that time.[10] Later in their careers, Len and Lee made sure that women were invited to speak at any conferences or meetings they were invited to themselves; an attitude which was quite pioneering in the 1970s and 1980s.[11]

World politics was always important to Len and Lee, and their social circle was entwined with their activism. 'The people who hung together, hung together,' as Lee puts it.[12] While at Caltech, they helped establish a chapter of the Federation of American Scientists, a liberal organisation founded by scientists who worked on the Manhattan Project with aims that included reducing nuclear weapons.[13] Len and Lee were inspired by stories of how Monod's lab hid Jewish scientists during the Second World War. Monod often reminded them that although it sounded noble or romantic in hindsight, it wasn't much fun at the time.

When Len received his draft notice for two years' military service he found a way to avoid it. 'With the Cold War escalating', Len has said, 'the United States Army wanted me to carry a rifle for my country. I preferred to carry a pipette.'[14] With Monod's help, Len and Lee moved to the National Institutes of Health in Bethesda because working at the US government research agency could count instead of army service. Lee thinks that later, Ronald Reagan's government restricted support for science in part to force scientists to work harder to fund their research, so that they would have less time for political activism.[15]

After two years at the National Institutes of Health, Josh Lederberg, who had just won a Nobel Prize for discovering that bacteria can exchange genetic material, recruited Len and Lee to Stanford. Here they began their work on the immune system and bumped up against the challenge of counting vast numbers of cells – with Len's bad eyesight. Specifically, Len and Lee wanted not just to count but to isolate different immune cells from one another. This would allow them to be characterised and their actions tested in well-controlled situations. In fact, across countless realms of biology, the challenge of obtaining pure populations of cells from all those present in any tissue, organ or blood sample was a real obstacle to progress. It was their immediate need, the strength of their partnership, their shared love of science and this wider challenge facing the scientific community that propelled Len and Lee to develop their breakthrough instrument. But also, a friend of theirs told me, what *really* spurred them on was something else – a deeply personal source of motivation.[16]

In November 1961, Lee and Len had their third child, a son after two daughters. Almost as soon as the baby was born he started turning blue. Without explaining anything, the nurses took him away. They knew the baby wasn't getting enough oxygen, but for a few hours nobody said anything to Lee. She was simply left to wonder where her baby was. Len, meanwhile, wasn't present at the birth, as was common for fathers at that time.

Talking to me in 2019, Lee can recall the moment she briefly held her son before he was taken away, and how there was something about his hips which didn't feel quite right.[17] Eventually, a doctor called Len to explain what was happening; then he told Lee. It wasn't good news: doctors thought their newborn baby would die within two or three months.

Their baby, Michael, had Down syndrome. Down syndrome is caused by the inheritance of an extra copy of chromosome 21, containing about 300 genes (from a total of around 23,000 in the human genome). This leads to changes in a baby's development. Michael had some of the most serious physical complications associated with Down syndrome, including a heart problem. He stopped breathing several times. Lee's grandmother argued that they should just take the baby home and he'd be fine, but Lee knew a chromosomal abnormality wouldn't go away by being at home. So Len and Lee never took Michael home. Looking back, Lee thinks that if she had taken Michael home, 'he probably would have died in my hands'.[18]

As it turned out, the doctor's prediction was wrong. Michael didn't die. But raising Michael was evidently going to be challenging. Lee was committed to science – 'It's a gift to be doing science' – and she didn't want to give it up.[19] So a paediatrician introduced Len and Lee to Barbara Jennings, a local woman who raised Michael along with a number of other children with developmental difficulties. 'It was selfish, if you like,' Lee says, 'because we had things we wanted to do ... [and] it would have been an intensive kind of upbringing, but on the other hand, he wasn't being put in a cot in an institute and I was thrilled to have someone share my child with me.'[20]

Had Lee known early during her pregnancy that she was carrying a child with Down syndrome, she says she would have had an abortion.[21] But at the time of Michael's birth, the genetic basis of Down syndrome had only recently been discovered, and it would be five years before any sort of prenatal test for Down syndrome would become available.[22] Ever since Michael's birth, Len and Lee had thought it vitally important to find ways to test a baby's health during pregnancy. They knew that a few cells from a developing foetus end up in the mother's bloodstream, and they thought that if a machine could isolate these rare cells, the baby's health could be checked.[23]

As it turns out, this goal was never realised, because there are too few foetal cells in a sample of maternal blood for this to work. But in 2008, another lab in Stanford, led by Stephen Quake (whom we met in Chapter Two as He Jiankui's postdoctoral adviser), did achieve something similar. His team showed that Down syndrome can be detected, not by analysis of foetal cells, but by analysis of foetal DNA, a little of which also turns up in the blood of pregnant women. Len and Lee celebrated Quake's work and helped it get published.[24] Even so, after Michael's birth in 1961, the notion of such a machine, and what a difference it might make to expectant parents, added a powerful motivation for Len's search for a way to count and isolate cells.

That search had taken him to Los Alamos, New Mexico. There, scientists had recently developed a machine to count and size radioactive particles. They were using this to assess radioactivity in the lungs of animals which had been sent up by balloon into the mushroom clouds of atomic bomb tests.[25] Len asked the Los Alamos scientists if they could modify their instrument to count labelled cells. They weren't keen on attempting to do so themselves because, they said, it wasn't part of their mission. So instead, Len persuaded them to give him the plans for their instrument. 'If science requires independent thinking,' he later wrote, 'it also depends on some very strange collaborations.'[26] Soon after Michael's birth, Len obtained the blueprint.

A simple coin sorter passes the money through a series of holes that decrease in size so that the largest coins are syphoned off first into one stack, the next-largest coins are diverted next, and so on. Cells, however, can't be sorted as easily as this, because many different kinds of cells have a similar size, and they can change their shape. Evidently, a machine far more sophisticated than a coin sorter is needed to separate minuscule cells.

The machine which Len and Lee developed is called a flow cytometer. Inside it, cells are forced to flow single file through beams of light.[27] There are different ways this can be done, but most commonly in modern machines, cells in one liquid are injected into a stream of a second liquid, called the sheath fluid. Through a naturally occurring process called hydrodynamic focusing, this pushes cells to travel in a fine line down the centre of the two liquids, like a coaxial cable. Sound waves are sometimes used to align the cells in an even tighter thread.

Crucially, the different types of cell in, say, a blood sample will already have been labelled with different fluorescent markers. Today, this is done using special protein molecules called monoclonal antibodies – a subject we will return to shortly. Initially, however, while building their instrument and in order to keep testing how well it worked, Len and Lee added dyes to cells in separate tubes and then mixed them together, to create samples with predetermined numbers of differently coloured cells.

Inside the instrument, laser beams precisely intercept the flowing cells, illuminating each one briefly as it cuts the light. The different fluorescent markers give off a different colour as the stream of cells passes through the laser light. Mirrors and lenses collect and focus the light onto detectors, each equipped with a coloured filter, which convert the light into pulses of electricity. Other detectors detect the quantity of light reflected by the cell, which provides information about its size and its internal complexity. In the early days, results were captured on Polaroid photos of an oscilloscope screen.[28] In a modern instrument, pulses of individual colours and the amount of reflected light are recorded from thousands of cells every second, and computer software displays the results.

All in all, it took seven years from the time Len obtained the plans from Los Alamos until he and Lee and their team had built an instrument that worked. Len could push people hard when needed – at one point telling engineers that the instrument would be useless unless it could count cells faster – but also Len and Lee tried to instil a family-like bond in their team.[29] Every Thursday night, they had everyone over to their house for scientific discussion over wine and beer, and they were well known in Stanford for hosting fun parties.[30] San Francisco in the 1960s and 1970s was a hub for hippy counterculture, and that vibe infused Len and Lee's lives and their labs.[31]

A book which Len and Lee always loved to share was *Life is with People* by Mark Zborowski and Elizabeth Herzog.[32] When I asked Lee why they loved this book, she said, 'Because life is with people,' and 'Creative science is with people too.'[33] The book describes the culture of the *shtetl*, the isolated small-town Jewish communities of Eastern Europe, the type where the musical *Fiddler on the Roof* is set. The *shtetl* culture emphasises tradition, human welfare and family life.[34] *Shtetls* were destroyed by the Holocaust, but for Lee there was something transmitted from *shtetl* culture to Jews with ancestry in Eastern Europe now in the USA, and this affected the tone of their laboratory: a 'recognition that buildings don't mean anything, places don't mean anything, money doesn't mean anything, but interactions amongst people – those will stay forever'.[35] This is not an aside to Len and Lee's science: it is at the core of what they believed, affecting how they ran their lab and their approach to everything.

The most important advance the team made together is that their machine did more than count cells. For many applications, counting is all that's needed. But to scrutinise the cells in any other way, the different cells needed to be sorted and separated. Their instrument which can sort cells – essentially a modified flow cytometer – is well known by the name Len gave it: the fluorescence-activated cell sorter, or FACS. However, the way it works was neither Len nor Lee's own idea: the principle had been developed by Richard Sweet, also at Stanford, as a way of controlling the position of ink droplets, needed to make an ink-jet printer.[36] The FACS instrument works just like a basic flow cytometer, but with a crucial adaptation. Before passing in front of the laser, the stream of cells passes through a small hole or nozzle which is vibrated so that droplets are produced, each containing a single cell.[37] The precise frequency needed to keep droplets falling off in the exact same place, each containing a single cell, is named after Sweet as 'the Sweet spot'. Just before each droplet buds off, the stream of cells is given an electric charge. A cycle of charging then de-charging the stream is timed so that each droplet, containing a single cell, is given its own particular electric charge, positive or negative, depending on which colours the cell gave off when hit with laser light.[38] A droplet containing one type of cell, labelled to shine green, might be positively charged, while those containing another type of cell, labelled red, would be given a negative charge.

Once charged, the droplets then fall between two electrical plates, one positive and one negative.[39] The droplets are attracted to their opposite charge: a positively charged droplet will bend its path towards the negatively charged plate, and vice versa. Any unwanted cells, which find themselves in uncharged droplets, will pass through the plates undisturbed. Tubes can then be positioned to catch cells deflected one way or the other and, in that way, different types of cells are sorted apart.

In 1969, their first paper describing a machine which could isolate, or enrich, cells of one type from a complex sample was published.[40] The instrument – dubbed 'the Whizzer' – was built in a basement of the medical school for around $14,000.[41] It's not that this came without precedent: this is a technology which combines many concepts – from physics, biology and engineering – and it would be wrong to give the impression that the Stanford team accomplished this out of the blue or entirely alone.[42] In the early to mid-1960s, for example, Louis Kamentsky and his colleagues, working at IBM, built a machine which could separate out cancer cells from normal cells.[43] It worked, but wasn't reliable enough to be used clinically.[44] IBM sent a version of this machine to Len in Stanford. Len said he never used its design, but he did re-use its parts.[45]

With the arrival of an atmospheric nuclear test ban treaty in 1963, scientists at Los Alamos no longer had to monitor radioactive fallout. So they too turned their attention to sorting cells. There, Mack Fulwyler built an instrument which could sort cells on the basis of their size.[46] Fulwyler also used Sweet's idea to achieve this.[47] Fulwyler's was an important precursor to Len and Lee's machine, although it wasn't able to sort cells according to any biological feature.[48] Fulwyler recalls that most scientists at Los Alamos – predominantly physicists and engineers – weren't especially supportive: they just couldn't see the point in scrutinising individual cells. They still thought 'in terms of taking a flask full of cells and grinding them up and measuring an average value of some characteristics'.[49]

Len and Lee knew the impact that sorting cells would have. They knew that unravelling the complexity of the immune system – or any part of the human body – required understanding the diverse characteristics and functions of its component cells. And they realised that if different types of cells could be sorted apart while still alive, each could be used in subsequent experiments, alone or in combination, to test their functions. Len is sometimes credited with leading development of this instrument solely, and scientific prizes have been given to him alone. But he always maintained that everything was shared with Lee. Together, Len has said, they 'turned a machine built as part of the world's most destructive enterprise into a powerful force for healing'.[50]

Even so, visionary as they were, it was impossible for them, or anyone else, to foresee just how powerful this type of instrument would actually become. It is every bit as vital to modern science as better-known technologies such as magnetic resonance imaging (MRI) or gene sequencing. Blood, tissue or tumour samples are nowadays routinely analysed by flow cytometry in labs and hospitals. As well as counting cell types, the presence of viruses or bacteria can be detected, or whether or not a person's immune cells are doing well or are impaired. The way in which different people respond to vaccines, for example, is also studied by flow cytometry. The same instrument, tweaked slightly, can be used to analyse the presence of genetic abnormalities, or to study the minuscule life forms that fill our oceans.

Although all cell sorters work with the same basic principles as Len and Lee used, today's instruments are incredibly sophisticated. The most lavish of them cost around $1 million, and use different lasers and detectors to scrutinise samples and isolate cells identified with multitudes of markers. It wasn't obvious at the outset that this would happen; Len approached several companies who said they weren't interested in pursuing the idea at all. It was Bernard 'Bernie' Shoor, from the US medical device and reagent company Becton Dickinson, who eventually saw the commercial opportunity. Shoor was visiting Stanford looking for technology to commercialise when Len told him that his cell-sorting machine was what he should pursue. Shoor wondered if his company might sell ten, or could they possibly sell thirty such instruments?[51] Len said he thought they might even sell a hundred.[52] Shoor took a punt on it and Becton Dickinson licensed Len and Lee's technology. With half-hearted commitment, they planned to sell instruments to order. Initially, they only promoted it to a select group of scientists they thought would be capable of putting it to good use.[53] Evidently, they grossly underestimated how big an asset they'd acquired. Demand almost immediately outstripped supply, and by the year 2000 around 30,000 flow cytometers were being used across the world's labs and hospitals.[54]

By taking on production of Len and Lee's machine, it can be argued that Shoor had built the world's first biotech company.[55] In 2018, the market for flow cytometry was about $3.7 billion.[56] Demand continues to grow. Recall that Len and Lee believed 'money doesn't mean anything'. They asked everyone named on their lab's patents to sign their royalties back to the lab, to continue the science.[57]

After 'the Whizzer', a relatively small community of academic and industrial labs carried on improving the technology, but from the mid-1970s onwards, the focus shifted to using, rather than developing, flow cytometry.[58] Our understanding of cellular diversity inside the human body – and all life on earth – promised to be transformed by this new-found technology. But even after Len and Lee's machine had been proven to work, another major advance was still needed.

Their instrument was useless without there being ways of labelling different kinds of cells. And at the time their instrument was first developed, only a few reagents were available to do this. There were dyes which could stain white blood cells separately from red blood cells, and dyes staining DNA could be used to signify the presence of cancer cells if they had an abnormal level of genetic content. Serum (the fluid part of blood) derived from animals could be used to mark some kinds of human cell (because the serum contained antibodies, which we'll turn to shortly).[59] But overall, a paucity in labelling reagents limited the use of Len and Lee's instrument at first. Thankfully, another revolutionary advance was just around the corner, which provided a way of marking cells very precisely.

In general, finding a way to label different types of cells is not as straightforward as it might seem. A typical neuron with multitudes of axons protruding out from the main cell body does look very different from, say, the flat, indented disc shape of a red blood cell. But these cells are exceptional in having a very particular shape; many others look similar under a normal microscope – small and round. What's more, every cell in a person's body contains the exact same set of genes (apart from sperm or egg cells, which have half the number of genes). What makes one type of cell differ from another is which of those genes have been 'switched on', which is what defines a cell's characteristics, abilities and function. Genes are the code for producing protein molecules, so when a gene is switched on, it simply means that the cell containing that gene will now produce the protein molecule encoded by that gene. To give an example: red blood cells have had genes switched on to produce haemoglobin, which binds and releases oxygen and gives red blood cells their ability to shuttle oxygen from our lungs to elsewhere in the body. (In fact, haemoglobin is made up of four protein molecules, two As and two Bs, encoded by separate genes.[60]) These genes are switched on in red blood cells – but not any other type of cell. To take another example: a type of immune cell called a T cell has at its surface a group of proteins assembled into what's called the T cell receptor, which is unique to that type of cell and crucial to its ability to detect infected or cancerous cells. Put simply, what each cell does in the body depends on the proteins the cell has. So to sort apart the different kinds of cells in, say, a sample of blood – which would include red blood cells, T cells, and countless others – Len and Lee needed a way to tag each cell's distinctive signature proteins.

In the autumn of 1976, Len and Lee arrived in Cambridge, England, to spend a sabbatical year working with César Milstein at the renowned Medical Research Council Laboratory of Molecular Biology. Shortly before they arrived, Milstein, together with a postdoctoral researcher in his lab, Georges Köhler, had also developed a new technology – not a machine, but a lab process – for which they would later, in 1984, win a Nobel Prize. What Milstein and Köhler had discovered was a way to produce a type of molecule that could attach itself to almost any other specific molecule of their choosing. Len and Lee brought the Cambridge lab's methods back to Stanford – not exactly with Milstein's blessing[61] – and arguably, that's when the cell-sorting revolution really began.

To follow this crucial development, we need to understand some of the basic science around the type of thing Milstein and Köhler had produced: antibodies. Antibodies are soluble protein molecules naturally secreted by some of the body's immune cells to stick to and neutralise infectious bacteria, viruses or other dangers.[62] The way antibodies are made is complex and one of the greatest wonders of the human body. Immune cells called B cells secrete antibodies and, importantly, each individual B cell produces just one version of antibody. All antibodies are roughly Y-shaped, but the antibody produced by each B cell has a unique shape – the variable region – at the two tips of its double-pronged end. This is the part of the antibody which sticks to its target molecule, which might be, for example, something on the outer coat of bacteria. The different shapes that are unique to each antibody mean each will stick only to its own specific target. But what's really amazing is this: B cells don't produce antibodies to be able to stick to germs as such. Instead, the top part of each antibody is created to have an almost random shape.[63] Then, as each B cell is created in the bone marrow, it is tested to see if the antibodies it produces happen to stick to anything that naturally occurs in the body: if it does, it is killed off or inactivated to avoid doing any damage.[64] That way, the only B cells allowed out from the bone marrow are those which make antibodies which would only stick to things *not* normally present.

On account of having around ten billion B cells, each of us has the ability to make something like ten billion differently shaped antibodies. Each of these can lock onto something which hasn't been in the body before. When an individual B cell does have the right antibody to lock onto something alien and dangerous, the B cell multiplies so that its useful antibody is produced in bulk. In this way, antibodies can be mass-produced against virtually anything alien to the body.[65] This is how our immune system can respond to germs which haven't been encountered before, including germs which have never even existed in the universe before.

This means that humans cannot naturally produce antibodies that target human proteins – which is what Len and Lee were after: reagents to tag human proteins. But a non-human animal could. And so to obtain such antibodies, mice (or other animals) are 'immunised' with a human protein – i.e. injected with a specially prepared version of it, which often includes other molecules that help trigger a strong immune response – and after a few days, the B cells that produce an antibody that attaches to that protein can be obtained from the animal's spleen (an organ which is abundant in B cells).

Normally, B cells can't survive for long outside the animal's body, but this is where Milstein and Köhler's Nobel prize-winning work comes into play. They fused antibody-producing B cells with cancer cells to create new cells that have the growth properties of a cancer cell, while still producing the antibody made by the original B cell. Köhler himself knew the idea sounded 'crazy',[66] but it worked. To fuse the cells together they used Sendai virus, first isolated in Japan in 1953, which has the ability to do this. And previous research in Milstein's lab had identified a mouse cancer cell which worked especially well in the fusion process.

Individual antibody-producing B cells were then isolated by diluting the cells sufficiently and adding a small amount of the resulting suspension into the minuscule indentations, or wells, of a rectangular plastic dish. The liquid from each well, containing the antibody, was then tested for its ability to bind to the desired target.[67] Any B cell found to produce an appropriate antibody was then cultured in a large flask to provide an almost limitless supply of it.[68] An antibody produced this way is called a *monoclonal* antibody, because a bulk amount of one antibody is produced by a culture of *cloned* cells. In other words, instead of a population of B cells making lots of different antibodies, every cell now produces the exact same antibody. Milstein had many hobbies, including cooking, and he referred to this process as a way to make antibodies à la carte.[69] Nowadays, monoclonal antibodies can be produced in other ways too, such as by transferring antibody-encoding genes directly into appropriate producer cells, but this is a detail.

It's hard to overstate the importance of what Milstein and Köhler achieved.[70] Medically, antibodies are used to kill cancer cells, to trigger immune activity against cancer cells, or to dampen immune responses in treating rheumatoid arthritis, multiple sclerosis or other autoimmune diseases. They are also used diagnostically, such as for detecting a hormone in a pregnancy test. The presence of certain antibodies is also the basis of us being able to test whether or not a person has been exposed to COVID-19, for example. Indeed, the current top ten of money-making medicines is dominated by monoclonal antibodies. Their production has become an enterprise worth nearly $100 billion every year.[71] But most importantly for our purposes, they are exactly what Len and Lee needed: dyes are easily attached to monoclonal antibodies, and they can thus be used to attach specific dyes to specific cells, meaning that a flow cytometer can then be used to count and isolate them. More than this, cells can be sorted not just according to whether or not they are tagged in a simple yes or no sense, but also in an exquisitely quantitative way, so that cells marked with, say, a low level of one antibody and an especially high level of another, can also be isolated. This means that subtle variations in cells can also be scrutinised, separated and studied.

Milstein and Köhler were focused on pursuing basic scientific knowledge of how antibodies are produced by B cells, not financial profit. But by never patenting their methods it's possible that they, and UK science at large, lost out. Margaret Thatcher, elected the British prime minister in 1979, openly blamed the scientists as well as their funders, the Medical Research Council, for not doing so.[72] Milstein, and many others, always felt the criticism unfair. An administrator at the Medical Research Council had in fact investigated pursuing a patent, but officials at the National Research and Development Corporation – set up in 1948 to help transfer technology from academia into companies – failed to trigger an application because they could not 'identify any immediate practical applications' of antibodies.[73]

At the time, a more pressing concern to Milstein and Köhler was a crisis in the lab. Just as their work was accepted for publication, their process for producing antibodies stopped working. Eventually, it was found that one of the solutions had been wrongly prepared.[74] In the meantime, a different attitude to patents was taken in the USA. The virologist Hilary Koprowski, director of a research institute in Philadelphia, filed patents for antibodies which his lab had developed. And here's the sting: to produce antibodies in his lab, he used cells which Milstein had sent him.[75] Those patents helped Koprowski co-found the company Centocor, an early and hugely successful biotechnology company in the USA.[76]

Throughout the rest of his career, Milstein spread his lab's methods and ideas widely and openly, with enormous benefits to humankind and, in the end, significant financial return to the UK.[77] A problem for using mouse-made antibodies as medicines is that the human body sees them as alien and mounts an immune response against them. To get around this, Gregory Winter, who worked in Cambridge with Milstein, found ways to make mouse antibodies more human-like (by switching parts of the mouse antibody genes for their human counterpart). Winter was quick to see the commercial potential of these so-called humanised antibodies, leading to patents, a spin-out company and several important medicines, which in turn raised a considerable amount of money for the UK.[78] Both Milstein and Köhler often emphasised how their work was a perfect example of how basic research can lead to big and important commercial enterprises.

Milstein retired in 1995, but in practice this only meant he stopped working on Saturdays.[79] He died aged seventy-four, in March 2002, after battling heart disease for many years. Just a few days earlier, he had submitted for publication a paper describing new details about how B cells produce antibodies.[80] Köhler also died of heart failure but, sadly, in March 1995 when he was only forty-eight.[81] In contrast to Milstein, Köhler had sought the possibility of taking early retirement at fifty.[82]

One well-known scientist has commented that Köhler's career might have been very ordinary had he not co-invented mono-clonal antibodies: 'Chances are that he would have blended in with the majority of unknown research scientists, doing unspec-tacular work.'[83] But this seems bizarre to me. Would we know the name Alexander Fleming if he had never discovered peni-cillin? Or Harper Lee if she'd never written *To Kill a Mockingbird*? Köhler did co-invent monoclonal antibodies. And spectacular doesn't even begin to cover it.

With monoclonal antibodies, studying cells by flow cytometry became relatively easy and much more precise, because specific human cells could be easily tagged, so the issue then became what this new tool should be applied to. Len and Lee, like many labs, devoted a significant part of the rest of their career to tackling HIV. 'Living in the San Francisco area in the early days of the AIDS epidemic was like living in a war zone,' Lee recalls, '[and] it was impossible to go to work in the laboratory without wondering whether something you were doing could be of help.'[84]

HIV can't replicate itself on its own – no virus can. For a virus to spread, it has to get inside the body's cells to hijack the cell's machinery for copying genes, normally used when a cell divides. Different viruses enter different types of cells, which is one reason why each is associated with specific symptoms. HIV gets inside immune cells called T cells by locking onto a human protein, named CD4, which some T cells have at their surface.[85] But as the virus moves in and out of these T cells, it destroys them, so that as the disease progresses, T cells displaying the CD4 protein decrease in number. In fact, due to a complex cascade of events – still not entirely understood – these T cells decrease in number by even more than those the virus kills directly.[86] This loss is important in two ways. First, a deficiency in T cell numbers is central to why people with AIDS commonly suffer from other infections. Secondly, the process can help diagnose whether or not a person has AIDS. From the early 1990s onwards, low numbers of these T cells – a low CD4 count – measured with flow cytometry, was widely taken as a defining characteristic of the disease. The pressing need for diagnosis of AIDS led to more rapid development of flow cytometers, and soon the instruments became cheaper and smaller so they could be more widely available, including in low-income countries.

Len and Lee weren't directly involved in developing the cocktail of drugs which eventually proved effective against HIV, now known as anti-retroviral therapy (or ART). But the tool they pioneered played a vital role every step of the way: bringing us a basic understanding of what happens to the immune system, clinically monitoring each infected person's state of health and, eventually, in assessing whether or not a treatment was working. Needless to say, AIDS remains a global health problem. Around 6,200 women aged fifteen to twenty-four, for example, become newly infected with HIV every week.[87] If a vaccine or other type of preventive medicine is ever going to arise, the path to its discovery will almost certainly involve Len and Lee's machine.

Flow cytometry has helped us understand all sorts of other diseases too. It allows us to study a biopsy from a tumour, for example, cell by cell. This has led to the idea that a person's tumour is not one but a million different diseases, on account of every tumour cell being subtly different. Each will have its own level of resistance or susceptibility to any one drug, which is why cancer patients are often given several drugs in combination. By analysing a person's tumour in such detail, treatments can be tailored accordingly.

As well as tackling disease, flow cytometry has transformed the way we see the human body in general. The average body contains something like 37 trillion cells, and flow cytometry has modernised the quest to understand what they all do. Many types of cells make up our immune system, for example – T cells, B cells and so on – but it has become clear, especially over the last few years, that these names are only coarse descriptors. Each individual cell is unique, containing a bit more or a bit less of each protein component found in that type of cell. One kind of immune cell which my own research team happens to study is called the Natural Killer cell. There are about a thousand of these immune cells in each drop of blood, and they are especially good at detecting and killing cells which have turned cancerous or have become infected with a virus. But not all Natural Killer cells are alike. One analysis has estimated that there are many thousands of variants of this immune cell in any one person.[88] In 2020, my own research lab carried out an analysis which suggested that variants of Natural Killer cells in blood could be organised into eight categories.[89] The different roles they have in the body aren't entirely understood, but some are likely to be especially adept at attacking particular kinds of virus, others are likely better at detecting cancer, and so on.[90] Other types of immune cells are just as varied, if not more so. Evidently, our component cells are as diverse as the human beings they make up, and understanding how such complex populations of cells work together, in this case to defend against disease, is a vital frontier. To penetrate the complexity, instruments which shuffle cells and liquid droplets must now be used in conjunction with computational analysis of the results. For this, the diversity of human cells must be translated into the language of algorithms, as follows.

Imagine a cell contains just two kinds of protein, X and Y. Every individual cell will have a specific amount of each of these two proteins. This can be represented as a point on a graph where the level of protein X becomes a position along the x-axis, and the level of protein Y its location along the y-axis. One cell may contain, for example, a high amount of protein X and a little of protein Y (which can be revealed by a flow cytometer showing that it stains with a high amount of one antibody and a low amount of another antibody). This individual cell can then be represented as a point placed far along the x-axis and a little way up the y-axis. In other words, the level of each protein becomes the cell's co-ordinates. As each cell takes up a position on the graph, those with similar levels of the two proteins – likely to be the same type of cell – appear as a cluster of points. If thousands or millions of cells are plotted in this way, the number of discrete clusters that emerge tells us how many types of cells there are. Also, the number of points within a cluster tells us how many cells there are of that type. The wonderful thing here is that this analysis can reveal how many kinds of cells are present in, say, a sample of blood or a tumour biopsy, without being guided in any way about which cells we might expect to find. This means that unexpected results can turn up. A cluster of data points might appear with unexpected properties – implicating the discovery of a new kind of cell.

Of course, cells don't just have two types of proteins: every cell contains something like 10,000 different types of protein. A modern version of Len and Lee's machine might be able to measure the levels of around thirty of them. The same principles are at work, but nowadays a high-end flow cytometer uses multiple lasers and detectors, as well as computational analysis to account for any overlap in the colour of light emitted from different dyes. We can't imagine cells represented on a graph with thirty axes, but a computer algorithm can handle the analysis in just the same way as it would with only two variables. In other words, cells can be organised on the basis of how much of thirty different proteins they each have, and different types of cells can be identified and scrutinised.

A recent advance in this technology is a variation dubbed mass cytometry, developed by Gary Nolan at Stanford University, who trained with Len and Lee.[91] In this, antibodies are labelled with different metal atoms rather than fluorescent dyes. The advantage is that many more labels can be used on the same sample, because metals can be more precisely discriminated by their mass and charge, compared to separating different dyes by their colours. With this, as many as one hundred different characteristics about individual cells can be measured. However, this type of analysis can go much further still.

An especially important method developed over the last decade – single-cell RNA sequencing – can measure the extent to which each cell is using each of the 20,000 human genes it has. To understand how this works, we need to consider how cells use genes. Roughly speaking, each gene encodes an instruction to make one type of protein molecule. When a gene is 'switched on', its DNA sequence is copied into another molecule, called messenger RNA, which then goes out from the cell's nucleus to trigger production of the corresponding protein. Sequencing a cell's messenger RNA molecules, therefore, tells us which genes are active, i.e. which genes have been switched on to produce protein.

The way this is achieved varies from one instrument to the next, and new instruments are being launched regularly, all taking advantage of the fact that sequencing genetic material has become easy and cheap. Early work in this area used Len and Lee's machine as the first step, to isolate the type of cell which will be investigated.[92] Nowadays, individual cells are often isolated inside water droplets that flow through channels of oil, inside a minuscule chip. This is a mind-boggling feat of engineering in its own right, and the outcome of a whole other research specialism called nano-fluidics. Then, the individual cells are destroyed inside these water droplets so that their messenger RNA contents come out and attach to a small bead which, in effect, is used to denote that all these RNA molecules come from the same cell. The messenger RNA is then sequenced. Most importantly here, the level at which a cell is using each of its 20,000 genes – called the cell's transcriptome – can then be analysed to create a 'map' of the cells. Similar cells are positioned close together, while cells using very different sets of genes are far apart. Algorithms to do this are borrowed from other fields of science, such as those used in analysing social networks. Then we get to spend days, if not years, mining the output, deciphering what the map means: how many types of cells there are, what defines their differences and then what they do in the body. To give an example, Len and Lee's machine can be used to isolate all the B cells in a sample of blood, then single-cell RNA sequencing can be used to chart every subtle variation of B cell that's present.

There are many difficulties in interpreting such data. For example, cells with different profiles of gene activity might be different kinds of cell, or they could be the same cell in a different state or situation, such as whether or not the cells have recently divided or not. Unravelling this requires the expertise of scientists from all different backgrounds: computer scientists, cell biologists and so on. This type of research is hard in all sorts of ways – from the science itself to the sociology of large teams – but the pay-off can be huge.

It certainly was for a consortium of twenty-nine scientists who set out to analyse the lining of a trachea, the 11cm-long tube which carries air from our throat into our lungs. One person involved in this work, Moshe Biton, takes great care to emphasise how no one person could ever do this type of project alone.[93] The Israel-born, US-based scientist Aviv Regev, who co-led the study, emphasises this point too: so many people were needed to make this happen.[94] To begin with, the team studied the trachea from mice. Six types of cell were already known to be present in a trachea, and they all turned up in the analysis. Subtle variations of each type were revealed – which was interesting but not a breakthrough. Far more importantly, a small number of cells didn't seem to correspond with anything ever seen before.[95] The first time the team came across these cells, they were looking at the analysis of a total of 300 cells and just three of them looked different from anything expected.[96] Had it been two cells, the team might have dismissed them as an outcome of noise in the data – but three strange cells warranted a closer look.[97] In the lab banter, they became known as the 'hot cells'.[98] They repeated the experiment several times, and it soon became clear that they really had stumbled upon a new type of cell in the trachea.

As it turned out, another team independently found the same thing. They learnt of each other's work by chance, when a person from one team went to a seminar by someone from the other team. 'It was one of those beautiful moments in science,' Biton recalls, 'when two groups found the same results separately.'[99] Both groups confirmed that these new cells exist in the human airways as well as in mice and, after meeting up, they agreed to publish their two papers together side by side.[100] Where once a primitive microscope, essentially little more than a magnifying glass, could reveal a new cell directly and viscerally, in the way Leeuwenhoek discovered sperm, today it is analysis on a computer screen which brings us that kind of revelation. But it's just as wonderful.

These new cells had not been noticed before simply because they are so rare, making up around 1 per cent of cells in the airway. But that doesn't mean they're unimportant. When the two teams looked in detail at which genes are uniquely used by these cells, they came across something astonishing. One of the genes especially active in these cells turned out to be *CFTR*. This gave their work a whole other level of meaning, because mutations in this gene cause cystic fibrosis.

Exactly *how* cystic fibrosis is caused by inheritance of a dysfunctional version of the *CFTR* gene has been a mystery ever since the link was discovered in 1989.[101] This is a complex disease, usually beginning in childhood, with symptoms often including lung infections and difficulty breathing. There are treatments but no cure. Now it seems possible that the key to understanding what goes wrong in cystic fibrosis could very well lie in working out what these newly discovered cells do, and what happens to these cells if the *CFTR* gene is defective. The research is ongoing. But already from this discovery, and other research using similar methods, there is the sense that our understanding of the body's cells is being transformed by this new, piercing combination of biology and computer science.[102]

And that's where many really game-changing discoveries are about to be made. Enter the Human Cell Atlas project.

In 2014, Regev started prefacing her research talks with an evangelical call to arms about a bold new project: the Human Cell Atlas.[103] A handful of other scientists were thinking along the same lines, and together they organised a meeting in London in October 2016, where a group of ninety-three scientists met and agreed that we need to define every cell in the human body. Their elevator pitch was to assemble something like Google Maps, but for the body: we know the countries and main cities, but now we need to map the streets and buildings. A year later, they had drafted a specific plan – to profile 100 million cells in the first instance, from different systems and organs, from different people across the globe.[104] Thousands of scientists, in over seventy countries, have since joined the project. The community to study the body's cells is in itself especially diverse, as it should be – something Regev is especially proud of.[105]

The scope and scale of the effort will, in fact, amount to something much more than a map. Identifying where cells are in the body, and which genes they're using, can also be analysed to reveal where and when different cells interact, which cells develop from other ones, and so on. By comparing samples from different people, we will gain a deeper understanding of how the body transitions from health to disease, for example, or from youth to old age. All of this derives from deep scrutiny of the human body's cells, which exploded with the invention of flow cytometry and continues now at a level which would have been utterly unimaginable at the time Len and Lee published 'the Whizzer'.

In many ways, this bold new ambition is also a direct descendant of the Human Genome Project. In April 2003, a finished version of the human genome sequence was announced and genetic research exploded. As a result, all sorts of genetic variations have been linked to an increased susceptibility to a specific illness. But crucially, genetic diseases manifest in the specific *cells* where that gene is normally used. The Human Cell Atlas project will bridge this gap, between abstract genetic sequences and the physicality of the human body. We've already seen one example of how important this is – the discovery of the cystic fibrosis gene being used by a new rare cell. Another example comes from what happens during pregnancy.

For many years, we have known that the immune system is intimately linked with pregnancy. For example, some combinations of immune system genes are slightly more frequent than would be expected by chance in couples having had three or more miscarriages.[106] We don't understand why this is, and working this out might be medically important in resolving problems in pregnancy. To tackle the issue, a consortium of scientists recently analysed around 70,000 cells from the placenta and lining of the womb from women who terminated their pregnancy between six and fourteen weeks.[107]

The placenta, as we've discussed in Chapter Two, is the organ where nutrients and gases pass back and forth between the mother and developing baby. It was once thought that the mother's immune system must be switched off in the lining of the womb where the placenta embeds, so that the placenta and the foetus isn't attacked as being alien, like an unmatched transplant, on account of half its genes coming from the father.[108] But this view turns out to be wrong – or too simple at the very least. In the womb, the activity of the mother's immune cells is somewhat lessened, presumably to prevent an adverse reaction against cells from the foetus, but the immune system is not switched off. Instead, Natural Killer cells – the immune cells we met earlier as being good at killing cancer cells – take on a completely different, more constructive, job in the womb: helping build the placenta. Indeed, Natural Killer cells from the womb turned out to have a particular profile of genes switched on, marking them as very different from their counterparts in blood. Analysis of 70,000 cells highlighted that all sorts of other immune cells are also important in the construction of a placenta. What they all do isn't yet clear – this is at the edge of knowledge. As one scientist put it, we're just at the beginning of being able to crack an 'immunological code of pregnancy'.[109]

Muzlifah 'Muzz' Haniffa, born in Malaysia to parents from India and now based in the UK, was one of the three women who led this project. When I ask how her journey began, she says, 'Dad wanted me to be a doctor from the moment I was born.'[110] Now, as a physician and scientist, she sees the body from two perspectives on an almost daily basis: a computational analysis of cells on a screen and patients who walk through the door. Stones and the arch they make. Right now, these two views don't easily mesh but, as our understanding deepens, they will. In the future, Haniffa thinks, the tools doctors use on a daily basis, like a stethoscope to listen to a person's lungs or a simple blood count, will be replaced by instruments which profile our body's cells.[111] Algorithms will analyse the results, clarify the problem and predict the best treatment. Other physicians agree with her – that this has to be what's coming.[112] We are already accustomed to the idea that our personal genetic information can be used to guide our health. But a quieter – almost secret – revolution is also under way, and it may have an even bigger impact on the future of healthcare: deep analytics of the human body's cells.

4 The Multi-coloured Brain

Cognition – reasoning, imagining, believing ... that's hard ... Where is it happening? How? If you had a choice, would you choose a mouse? Would you choose optics? Would you spend your time looking at a brain while you poke it with a laser?

Tom Stoppard, *The Hard Problem*

In 1873, the Italian biologist Camillo Golgi discovered that a combination of chemicals – silver nitrate and potassium dichromate – could be used to highlight the outside edge of cells, making them visible under a microscope. Fifteen years later, the Spanish scientist Santiago Ramón y Cajal used these chemicals to stain slices of brain and made a ground-breaking discovery. At the time, it wasn't clear what the brain was made of. Golgi always maintained that it was made up from a continuous network of fibres. But Ramón y Cajal realised this wasn't true. He saw that the brain was made up of separate cells – neurons. And where two individual neurons connect, he realised that a minuscule gap exists between the edges of the cells – a synapse – which we now know are the junctions where chemical and electrical signals are transmitted from one cell to another. Golgi tried to refute Ramón y Cajal's claims, and their disagreements soon became personal. Eventually, Ramón y Cajal withdrew from the argument – and was proved right – but Golgi remained bitter and continued to complain about Ramón y Cajal in the halls and lobbies of scientific meetings.[1] They met only once, when they received a Nobel Prize together in 1906. Even on this occasion, Golgi's speech attacked 'the neuron theory', suggesting it was out of favour – which it wasn't.[2] Their feud has become legendary because their work was so important. The discovery that the brain is made up of separate cells which communicate with each other across synapses is the foundation of our understanding of this organ, and opened up the therapeutic possibilities of brain surgery and neurological medicines.

Now, over a century later, synapses can be studied in exquisite detail. The protein molecules which accumulate there can each be isolated and examined at the scale of individual atoms. In this way, for example, we can see precisely how the drug LSD locks onto the particular receptor proteins that detect serotonin, a chemical neurotransmitter that plays a key role in all sorts of brain activity.[3] Or how an anti-psychotic drug connects to the receptor for dopamine, another key neurotransmitter, in an atomic-scale jigsaw fit.[4] This level of detail can – at least in principle, if not yet realised in practice – help us design new medicines which lock onto these targets in the brain even more tightly, hopefully reducing the likelihood of side-effects. But this sort of atomic-scale view doesn't reveal much about how the brain *really* works. It's simply too close up, like trying to understand the *Mona Lisa* by analysing the chemistry of Leonardo's paint.

A broader view is available with the help of functional magnetic resonance imaging, or fMRI.[5] With this we can form a picture of a living subject's brain activity. A person lies down with their head in the hole of a large multi-million-dollar doughnut-shaped machine, which uses a strong magnetic field to detect the flow of oxygen-rich blood within the brain – an indicator of activity. This works because haemoglobin – the component of red blood cells that carries oxygen and releases it where needed – has subtly different magnetic properties depending on whether or not oxygen is bound to it.[6] This type of scan can detect, for example, which parts of a brain have become affected by a stroke or trauma. It is also one of the main technologies used to generate the familiar picture of the brain with a part of it 'lit up' in response to a particular stimulus or experience. This high-tech instrument has been used in count-less experiments, including one that set out to understand that most vital of contemporary debates: Coke v. Pepsi.

Chemically, these two drinks are very similar, yet people often have a strong preference for one over the other. To understand why, scientists scanned the brain activity of people as they drank the two drinks. Amazingly, the parts of their brain that were active when tasting the drink changed depending not on which brand they were drinking but on whether or not they *knew* which brand they were drinking.[7] Being told beforehand which brand they were tasting was seen to increase activity in parts of the brain associated with memories and cognitive control. Although causation is hard to prove, people enjoyed drinking Coke more when they knew it was Coke. Similarly, another study found that people preferred the taste of a wine when they had been told it was expensive.[8] A part of the brain involved in experiencing pleasantness was more active when they thought the wine cost more. Evidently, our preferences for drinks are not solely based on how they taste in the mouth.

In fact, while it feels as though we observe and witness the world around us, everything we experience is actually created in our brains. Light, for example, is an oscillating wave of electric and magnetic fields, a physical thing and a form of energy. It doesn't really have a colour. Rather, our brain interprets the frequency at which it oscillates as the sensation of colour. The things we see in the outside world obviously exist separately from us, but the drama of a sunset, the spectacle of a rainbow and the way we picture another person are all created in our own heads. Beauty lies in the brain of the beholder. Likewise, Pepsi, Coke and different wines are all just mixtures of mol-ecules; their flavours and our preferences are created in our brains.[9] As Morpheus explained to Neo in the 1999 science fiction movie *The Matrix*: 'Real is simply electrical signals inter-preted by your brain.'

Studies using fMRI reveal interesting things about which parts of our brain are involved in our behaviours and feelings.[10] But they too don't get at how the brain *really* works. That's because fMRI and other medical imaging techniques can't see the activity of individual neurons. In practice, their maps of brain activity are the outcome of a sophisticated statistical analysis which treats the brain as a series of cubes, or voxels, each about one cubic millimetre in size, and each containing around a million neurons. They can show if an area of the brain displays heightened activity, but there's no telling which specific neurons are firing or what the consequences are. In other words, while the atomic-scale view is too close up, with fMRI scans we have zoomed out too much. It's like trying to understand the *Mona Lisa* by analysing how many people go and see it. That there is more activity around the *Mona Lisa* than other paintings in the Louvre says something about the painting's importance, but nothing about the thing itself.

Most scientists agree that the key to understanding the brain lies in understanding its circuitry – which neurons are connected to which other neurons. But this has proved exceptionally difficult to study for all sorts of reasons. The paramount problem is that the thin protrusions that connect neurons to each other are notoriously difficult to track, and there are an unimaginable number of them. A human brain is made up of 86 billion neurons, and each has multitudes of long, thin strands protruding from its main cell body: dendrites for receiving signals and an axon for sending them out.[11] When a message moves from one neuron to another, an electric signal travels the length of one cell's axon and triggers the release of neurotransmitters into the synapse at its end, to be detected by the receptor proteins protruding from the surface of the receiving neuron. Altogether the 86 billion neurons are connected by around 100 trillion synapses, each allowing messages to move from one cell to another. All sorts of messages can be sent across a synapse in a chemical language we barely understand. There are well over a hundred different types of neurotransmitter, and even more variation in the receptors which detect them. As well as this, some synapses also allow electrical signals to transmit from one cell to another, which only adds to the complexity.

There is also a vast diversity in neurons themselves. Purkinje neurons, for example, named after the Czech scientist who discovered them in 1837, have one axon and an especially dense branching tree of dendrites, while bipolar neurons, often involved in the transmission of senses, have one axon and one dendrite. But such textbook-level descriptions belie neurons' actual variability. In truth, we don't even know how many types of neuron there are.[12] And perhaps surprisingly, the human brain contains far more than just neurons. Neurons aren't even the most common type of brain cell – which are, in fact, glial cells. There are around 100 billion glial cells in a human brain. They were once thought to do almost nothing, but are now known to be involved in all sorts of activities, including forming and adapting neural connections.[13] They too come in all sorts of varieties, and their importance is only beginning to be appreciated.[14]

In short, inside your head is the most complicated object we know of in the universe. A small object responsible for all art and culture, the creation of money and bombs, everything humankind has ever done to the planet and the extinction of countless other species – not to mention our personal feelings, memories, dreams and relationships. And perhaps most mysteriously, our sense of self and the experience of making choices. For medicine, these cells and the way they are networked contain the secrets to understanding Alzheimer's, Parkinson's, epilepsy, schizophrenia, autism, depression and more. The scale of the problem – to understand the human brain – and its importance are second to none in all biology and perhaps all science.

When Ramón y Cajal discovered the synapse, he did so by using Golgi's stain to label cells sparsely, so that the edge of only one in every few hundred cells was highlighted. That way, he could pick out the shape of individual neurons. But to get a sense of the overall circuitry of the brain, we need to see *all* the neurons, rather than just a few. Yet, if all the cells in the brain were stained with the chemicals Ramón y Cajal used, the entire thing would appear the same black-brown colour and individual cells, let alone their protrusions, would be impossible to pick out. What we need is a way to highlight all the neurons, but each one separately. Enter Brainbow – a technology which takes seeing inside the brain to a whole new level.

Jeff Lichtman had been working towards the invention of Brainbow for decades, if not his whole life. His father was a physician, and at home there was always a microscope to play with.[15] This, he says, gave him a big advantage – that he never had any fear of playing with lab technology. Whereas the pioneers we met in Chapter One, who trained in physics or maths, thought deeply about optics and light pathways, Lichtman was happy just to build a microscope and tinker with it. Early in his career, his fiddling led to the ownership of several patents on microscope designs. More importantly, Lichtman trained as a physician and always had a specific biological goal in mind: to understand the brain. In his 1980 PhD thesis, he had noted how important it was to find a way of seeing everything in a brain, and later his lab team at Harvard made several attempts to see neurons. But it took until 2005 before he found something that looked as if it could really work.[16]

The big idea of Brainbow is to reveal all the neurons in the brain by colouring all of them in, each with a different colour. In the same way that a TV or computer screen mixes three colours – red, green and blue – to create all the colours of the rainbow on-screen, Brainbow works by attaching to each separate neuron a different amount of red, green and blue fluorescent proteins. The amounts assigned are random, which makes it almost certain that every neuron will end up being a different colour to adjacent neurons. That way, at least in principle, every cell can be picked out separately.[17] To achieve this, some genetic trickery is used.

In Lichtman's lab, the genes which encode for differently coloured fluorescent proteins – green, blue and red – were added into the genome of a mouse embryo. (The green and blue proteins were variants of the fluorescent protein that Shimomura first isolated from jellyfish in 1962, while the red protein was discovered in Moscow in 1999 and came from coral.[18]) Attached to them were small stretches of DNA that ensured the coloured proteins would only be produced in neurons. But the most important part of the trick was this: the inserted genes were added as multiple copies of a cassette, each of which contained three genes for the three colours, red, green and blue. Within each cassette, the colour-encoding genes were flanked by a small piece of DNA which is the target of an enzyme that can remove or inactivate genes. Then, while the mouse brain developed, the system was designed so that one randomly selected colour from each cassette was left intact, while the two others would be removed or switched off. Only those genes left intact would actually produce fluorescent proteins. So as each cell had multiple copies of the cassette, each having one randomly selected gene kept intact, each cell ends up with a different set of coloured proteins overall. For example, one neuron might end up having two copies of the gene producing red protein and one copy of the blue-producing gene, giving it a reddish-purple appearance overall. Other neurons, meanwhile, will end up with other combinations of genes and take on different colours.[19] Statistically, because something like a hundred colours could be easily obtained, the chances of any two adjacent neurons ending up the exact same colour was small. As the co-leader of the project, Joshua Sanes, put it, the genes were set up to work like a slot machine: 'One time it will come up cherry, orange, lemon. And another time it will be lemon, lemon, lemon.'[20]

One day in 2005, Lichtman and Jean Livet, a young French researcher in Lichtman's lab who did much of the work described here, placed a sliver of the cerebral cortex of this genetically modified mouse under the lens of their state-of-the-art microscope.[21] A computer screen connected to the instrument revealed each colour in turn. The red proteins were revealed first. As a laser beam scanned the sample, splashes of red built up on the computer screen. Lichtman recalls being thrilled that the image showed some cells bright red, others with only a little red, and some dark areas where, by chance, cells had produced very little red protein or none at all. Then, like a curtain coming down the screen, the blue proteins appeared. Once again, some cells lit up brightly, others dimly and some not at all. Where the blue and red overlaid, cells turned various shades of purple. The green layer appeared next and finally, before their eyes, was a multi-coloured slice of brain. 'God, that is unbelievable,' Lichtman recalls saying, thinking it the most amazing image he'd ever seen.[22] When the Brainbow pictures were officially published in 2007, they appeared all over the international press; as captivating as Hubble telescope images, except here was the cosmos in a brain.

This was just the first version of Brainbow, and the process has since been improved.[23] Identifying neurons in Brainbow pictures relies on a computational analysis being able to pick out different colours in 'noisy' microscope pictures, which is only practical for around 100 or so different shades. One way of increasing this number, so that more cells can be identified, is to modify the genes so that the coloured proteins only appear in specific places within the cell. For example, one set of fluorescent proteins can be located to a cell's surface with another targeting its innards. In effect, this gives each cell a separate colour for its outline and its interior. Yet another set of proteins can also be used to mark specific structures within cells, such as the energy-producing mitochondria. In other words, genes can become art tools; a simple version of Microsoft Paint made real for brain cells.

The big question scientists face today is no longer whether the brain is made up of a continuous network or of separate cells – the issue Ramón y Cajal and Golgi fought over – but is something like this: what is the difference between the brain of someone who can, say, ride a bike when compared to that of someone who can't? Or as Lichtman puts it, 'What would bicycle riding look like? How much would it weigh? Where is it?' 'Questions like this,' he says, 'can't be answered without getting down and dirty with the deep mysteries of the wiring diagram of the nervous system.'[24] Indeed, the wiring diagram of the brain – a map that shows which neurons are connected to which other neurons – is thought to be so important that a new word has been made up to describe it: the connectome.[25] A Google search for the word 'connectome' in 2005 would have brought up about ten results, many of which assumed a different word was meant.[26] Now the same search brings up over a million hits. Sebastian Seung, a computer scientist at Princeton, says 'you are your connectome'.[27] He thinks that mapping 'an entire human connectome is one of the greatest technological challenges of all time'.[28]

It's possible that a connectome isn't enough to reveal how something like 'riding a bike' manifests itself in the brain. Subtleties like the strength of signals at each synapse, for example, or their dynamics, might be crucial as well. Indeed, the wiring diagram of a person's brain may well change day by day, to some extent. But Lichtman thinks a lot of what the brain does is not so subtle. The brain changes dramatically, especially during infancy and early childhood. Synapses are overproduced at first and then pruned as we grow up. A one-year-old child has a brain with about twice as many synapses as an adult.[29] It's as if we begin life with a brain which wires up indiscriminately and then simplifies down to what's actually needed. In other words, the removal of a large number of synapses tunes the brain to suit our experiences. This suggests that at least some of our experiences shape the wiring diagram of the brain significantly, not subtly – which fits the idea that 'riding a bike' will be somewhere in a brain's connectome.

Even if a brain doesn't hold 'riding a bike' in any simple way, obtaining the connectome is almost certainly a good first step. Lichtman likens this to the way in which obtaining the complete sequence of a human genome was so vital to genetics. At the outset of the Human Genome Project, we didn't even know how many human genes there were. Now, the sequence of the human genome has become the basis for tackling all sorts of important questions, including how genetic variations affect health and disease (which we'll come back to in the final chapter). Matthew Cobb, author of *The Idea of the*

Brain, agrees with Lichtman: although we don't know exactly what a connectome will tell us, it's a good first step – 'It's a level of detail we're going to need.'[30]

Obtaining the connectome may help reveal principles and ideas which apply elsewhere in the body too. The brain is special in many ways, but all sorts of body systems involve networks of cells, and they too are 'wired' to behave in a co-ordinated fashion. For example, we understand various parts of the immune system – which cell is good at engulfing bacteria, which is better at detecting a viral infection – but there are major gaps in our knowledge of how the system works as a whole. The problem is that the immune system involves billions of cells moving around the body, through the blood and tissues, setting up untold numbers of brief connections with other cells, which makes the immune system's 'wiring diagram' very hard to capture. At least, as Lichtman says, the brain has a wiring diagram we can get at.[31]

Despite all this promise, the scientific power of Brainbow hasn't yet matched its visual glory. 'We didn't learn as much as we hoped,' Lichtman says.[32] Even in a thin slice of brain, there were just so many thousands of branched dendrites and axons, all overlapping and tangled, that it proved impossible to trace them. At a distance, the images were wonderful, but close up, the coloured lines soaked into one another. The culprit was the microscope. It just wasn't powerful enough to resolve so many extremely fine details in a dense network. Lichtman's team have tried to use super-resolution microscopes – those we met in Chapter One – which improved things, but not enough.[33]

In the quest for understanding the brain or even just defining its connectome, Brainbow has turned out to be a way station rather than a destination. In the future, we may very well return to it: what if, for example, the colours and hues aren't just random, but turn out to be indicative of something significant, like a neuron's activity or history?[34] In the meantime, another method for colouring in cells and picking out synapses has proven more effective.

As we saw in Chapter One, the wavelength of light puts a limit on what can be seen – on how much we can 'zoom in' – with a regular light microscope, but we have been able to get around this with the invention of other kinds of microscope. Instead of light, an electron microscope uses beams of electrons, which, according to the maths of quantum physics, can be described as having a wavelength around a thousand times smaller than light does.[35] The detail of this is hard to grasp, even for experts, but what it means is that an electron microscope can magnify the structures of cells much better than a light microscope. When it comes to tracing neurons, though, there is a crucial problem: an electron microscope can only picture an object's two-dimensional surface. While light can penetrate samples to some extent, depending on their transparency, electrons cannot. To use the power of an electron microscope to trace the protrusions from neurons, which twist and turn in three dimensions, a trick is needed.

Sometimes in science a simple idea works fine. Winfried Denk at the Max Planck Institute in Heidelberg – who trained in physics and whose passion is designing new laboratory techniques[36] – had the idea to install an automated cutting device inside the sample chamber of an electron microscope. The piece of brain being examined would be embedded in a plastic resin to make it easy to cut. Once a picture of its surface had been taken, an extremely sharp knife would then shave off its topmost layer, while holding it in exactly the same position, allowing a picture to be taken of the layer just beneath the surface. Repeating this process would result in a series of images that collectively reveal the whole sample in three dimensions.[37] The core technique itself wasn't new – the electron microscope was invented in 1931 – but here was an automated way of using an electron microscope to capture detail *throughout* a small piece of brain.[38] This way, the winding paths of dendrites and axons could be revealed.

To say that the knife used in this technique is extremely sharp is something of an understatement. Paper is around a tenth of a millimetre or 100 microns thick, and can easily cut a finger.

The edge of a typical kitchen knife is around three times sharper, and a razor blade is sharper still, thinned down to less than one micron at its edge.[39] This knife, however, is made from gem-quality diamond. Unlike a razor blade, which appears jagged and pocked when seen close up, the blade of a diamond knife is perfectly uniform. Its edge is about 0.002 microns thick, which is the width of about twelve carbon atoms. It can cut a single red blood cell into 300 slices. And unlike a razor blade, it doesn't easily blunt. Evidently, diamond isn't all sparkles; it can slice the body to reveal its secrets.

The sharpness of the knife is especially important because, with Denk's technique, the resolution at which neurons can be pictured comes down to how thinly the brain can be cut. Together with computer expert Sebastian Seung, Denk used his method to study a small fragment of retina taken from the back of a mouse eye. A retina doesn't merely capture light, nor does it send a 'picture' directly to the brain. Instead, a network of neurons in the retina separate, organise and filter information before passing signals on to the brain. Seung and Denk, and their lab teams, analysed a small cube of retina, about a tenth of a millimetre across, containing about 950 neurons connected by half a million synapses.[40] It took a month to acquire all the images and four years to analyse them. It turned out to be exceptionally hard for any computer algorithm to trace the dendrites and axons from one picture to the next, especially as they could branch at any point. So in the end, the team enrolled about 200 undergraduates to trace the neurons manually.

Realising that even this took far too long, Seung's team went on to develop an online game to crowd-source the effort.[41] Players of the game – dubbed Eyewire – traced neurons to earn points based on how many images they analysed, how long they took, and how much of their analysis agreed with others'. Over 265,000 people enrolled online, and the keenest spent fifty hours a week on it. By converting this laborious task into a game, a 3D view of multitudes of neurons was created and uploaded for anyone to browse.[42] There's still a long way to go in understanding how the retina processes information, but Denk and

Seung's work has shown that complex networks of neurons can be mapped with the unlikely combination of an electron microscope, a diamond knife and a computer game.

After Brainbow, Lichtman's team also turned to using an electron microscope, but they took a different approach. Where Denk's method discarded the shaved-off slices of brain, Lichtman's lab stored and took pictures of them instead of the block that remains. To build a machine which could do this, and use it to see all the synapses in what amounts to just a small crumb of mouse brain, took them six years.[43]

What Lichtman's team built somewhat resembles a movie projector spooling tape around a series of wheels. A small piece of mouse brain, again embedded in a solid resin, is moved up and down against the blade of a diamond knife in the manner of an automatic cheese slicer, with the thin slices falling onto the moving tape, which then feeds them in turn into an electron microscope. About 1,000 slices of brain can be cut every twenty-four hours in this way, but the speed at which a normal electron microscope operates is far slower: to achieve a full picture of even one cubic millimetre of brain would take around seventeen years.[44] To speed up the process, Lichtman's team used a new prototype of an electron microscope which scanned samples with sixty-one electron beams instead of one. They then used a computer to stack all the images together, so that the protrusions of neurons could be traced and coloured in digitally. In the end, the minuscule piece of brain analysed in this way was around a million times smaller than a cubic millimetre; far too small to contain even one whole neuron.[45] It was 'a lot of work for the complete rendering of almost nothing', Lichtman quips.[46] And yet, this little bit of mouse brain still contained the protrusions from multitudes of neurons – 1,407 axons and 193 dendrites to be precise, connected by around 1,700 synapses.

The team were surprised by how often the same axon and dendrite would connect together at different places. More importantly, perhaps, the images proved that synapses don't just form between any two protrusions that happen to be close to one another. One axon which ran through the length of the sample connected to only a small number of the other neurons nearby. This means that some other factor, still unknown to us, must cause neurons to connect, independent of their location.

New methods for imaging the brain are still being developed, and as this work continues more features of the brain are certain to be discovered. As Lichtman says: 'We have never had tools, until now, to see [the brain] at its full resolution. [And] we have just scratched the surface!'[47] Yet what this work also highlights is just how enormous the task of capturing a connectome is. Lichtman has estimated that a complete wiring diagram of the human brain needs about as much data as all the digital content held in the world today.[48] So 'It's not going to happen anytime soon,' he says.[49]

Obtaining the connectome of a simpler animal is easier – and was achieved decades ago. In 1986, a team led by the visionary South African biologist Sydney Brenner reported the connectome – although they didn't call it that, because the word hadn't been invented yet – of a small roundworm. Although the worm doesn't have a brain as such, it still has a nervous system which controls its behaviour. The effort to map it took well over a decade because none of the process was automated.[50] The worm's body, about a millimetre long, was thinly cut, every slice pictured with an electron microscope, and every neuron traced by hand. The outcome was 340 pages, describing 302 neurons.[51] The final analysis included an amalgamation of pictures from several specimens of the same sex. Only recently, in 2019, were complete descriptions of the nervous system of both worm sexes finally reported.[52] This recent analysis included all sorts of nuances, describing not only which neurons are connected, but also the physical size of each synapse, which is thought to relate to how strong each connection is.

From decades of worm neuroscience, we now know which neurons are important for a worm to sense its local temperature, allowing it to move away from anywhere too hot or too cold.[53] Likewise, we also know which neurons are involved in responding to touch, observable when a light pressure on its head makes it move backwards.[54] But there are differences between the nervous systems of worms of different sexes, and we don't yet know what they mean. More generally, how much variation there is between the connectomes of individual worms hasn't been looked at. Do worms have a personality somewhere in their connectome? Similarly, we do not know the extent to which a worm's life experience affects its nervous system. In fact, we don't even really know how best to represent a worm's connectome.[55] As one scientist commented: 'Depicted graphically, the new [worm] connectomes don't obviously resemble artificial neural networks or the wiring schematics of simple electronic devices; they look more like the cobwebs that lurk at the back of the broom cupboard.'[56]

Evidently, once a connectome is acquired, the hard work really begins. For so many parts of a body, lots can be learnt from just how it looks. That the human heart has four chambers, for example, is a clue to the fact that blood is pumped in two circulatory loops, first to the lungs, then out to the rest of your body. Even on a minuscule scale, the famous double-helix shape of genetic material is not merely ornamental: its discovery helped us understand that each of the two strands acts as a template for copying genes when cells divide. The brain, however, is an altogether different proposition. Even if a complete wiring diagram of the human brain was in hand, and even if the wiring diagram did contain 'riding a bike', we can't understand how it works just by looking at it.

To understand the neurological origins of behaviour, memories and emotions, we must be able to probe the brain as well as map it; we need tools that allow us to control or manipulate the activity of neurons and test the consequences. Enter the technology which, in 2010, the journal *Science* named as one of the breakthroughs of the decade: optogenetics.[57]

Most of us would consider the scientific mission to understand a brain as a serious and noble endeavour. Many of us may well consider understanding pond algae less of a priority – perhaps nothing more than a pursuit of esoteric details about nature, of which there are an infinite number. And yet in its current form optogenetics, a tool to study the brain, owes its existence to the science of algae. This is one of the most magical things about science and being a scientist: anything might turn out to be revolutionary.

For decades, a small community of scientists was trying to understand how the single-celled green algae found in ponds move towards a source of light.[58] In other words, how algae cells 'see'. An answer came in 2003 when a team based in Frankfurt, Germany, discovered a protein molecule in algal cells which converts light into electricity, triggering their activity.[59] The protein they discovered sits on the surface of an algal cell; when hit with light, it changes its shape to form a small hole. This opens a portal through which charged atoms move, causing a cascade of events which culminate in whip-like structures that protrude from the cell beating so that the cell moves breast-stroke style. It's this protein – which uses light as a cue to switch on the activity of a cell – that lies behind optogenetics. By genetically modifying neurons – or indeed any other type of cell – to produce this protein, we can switch their activity on with light.

This is not to say that this discovery in algae led to the development of optogenetics – the concept of which had already been building for some time – but that the algal protein provided a way to bring the idea into reality. In fact, in 1999 Francis Crick, who had co-discovered the double-helix shape of DNA, guessed that optogenetics might be possible. In a lecture exploring the techniques he thought were needed for better understanding the brain, he said: 'This seems rather far-fetched, but it is conceivable that molecular biologists could engineer a particular cell type to be sensitive to light.'[60] Five years later and it was two young researchers, Edward Boyden and Karl Deisseroth, who used the algal protein to fulfil that vision.

They both worked in Richard Tsien's lab at Stanford University, and liked to brainstorm late at night about technologies which might be able to control neurons. Boyden had trained as a physicist, Deisseroth as a physician, and their different perspectives drove them to consider all sorts of possibilities. They considered using magnetic beads, calculated that it might be possible, but realised it would be difficult to implement. In time, they realised that a better option would be a light-driven channel – the sort of protein algae use. They were encouraged by the fact that, in 2002, a team in New York used three genes isolated from fruit flies to switch on rat neurons with light.[61] Pioneering as this was, the system was complex – lots of components were needed to make the cells responsive to light – and it took some time for the neurons to react.[62] The algal protein, on the other hand, converts light into an electrical signal in a single step on its own.

In March 2004, Deisseroth sent an email to one of the scientists involved in the algae research, Georg Nagel, requesting the gene which encodes the light-switchable protein. Deisseroth soon began leading his own lab in Stanford, and he and his first PhD student, Feng Zhang, worked out the right conditions to get the algae gene into neurons, with the help of a virus.[63] Chance played its role: Zhang met Deisseroth only because he went looking for the person who used to occupy Deisseroth's office.[64] Deisseroth persuaded him to stay.

At around 1 a.m. on 4 August 2004, it was Boyden who carried out the crucial experiment. The very first neuron he tested gave an electrical signal in response to blue light. After a night of experiments, he told Deisseroth the good news by email. Deisseroth replied, 'This is great!!!!'[65] Crick – who had the vision – would never know. Aged eighty-eight, he had died a week earlier.

Despite the impact this research would eventually have, the top scientific journals refused to publish it at the time. *Science*, for example, argued that the team hadn't made any actual discovery beyond demonstrating the method. Eventually, in August 2005, their work was published in another journal.[66] Within months, other labs reported similar observations. Evidently, many teams had been chasing the same goal at the same time, using different strategies.[67] In 2006, the word 'optogenetics' was coined – which in itself connotes an exciting-sounding combination of optics and genetics.[68] But *Science* wasn't entirely wrong: there was still the formidable task of using optogenetics to provide new information about the brain.

Deisseroth spent the next few years getting up at 4 or 5 a.m. and going to bed at 1 or 2 a.m., in his search for a way to make optogenetics work in a living animal.[69] His wife, Michelle Monje, is also a neuroscientist at Stanford and leads a team seeking new therapies for brain cancer, particularly for children. She was once a competitive figure skater, and at the age of thirteen created a figure-skating programme for children with Down syndrome.[70] Monje and Deisseroth have four children together and for both of them, juggling everything hasn't always been easy.[71]

There were two big problems Deisseroth had to solve. First, he had to get the light-switchable algae protein into brain neurons, and secondly, he had to find a way for light to penetrate the brain to switch the neurons on. The first problem was solved, once again, with some genetic trickery. A short genetic instruction was added to the basic algae gene, packaged into a virus, and then injected into a living, fully developed mouse. This allowed the algae protein to be produced in a specific type of brain neuron. The second problem was solved with a fibre optic. One end of the fibre was attached to a laser, and the other was surgically inserted into the animal's brain. The mouse remained free to move, but with a thin optical fibre attached to its head.[72]

The first experiments targeted a part of the mouse brain known to be active when an animal wakes up after sleeping.[73] Having rigged everything up, a member of the team could switch on the laser while the mouse slept. After ten seconds of light passing into the animal's brain, the mouse gave a sudden jerk. Then fell back asleep. The seeming insignificance of this brief twitch belied the importance of what it meant: that optogenetics could switch on neurons in the brain of a living mammal and change its behaviour.

There are other ways to switch on neurons directly in an animal's brain, such as electrical stimulation. But this affects a general region of the brain rather than specific neurons. Drugs can also affect brain activity, but the timing of their effects is hard to control. The crucial advance here was precision: optogenetics could switch on a specific type of neuron, in one region of the brain, at any given moment.

When a reporter from the *New York Times* was due to visit the lab, Deisseroth asked Zhang if he could set up an experiment which might look especially impressive. So Zhang set things up to stimulate the motor cortex part of the brain, and the outcome was indeed dramatic. The mouse was sniffing the corner of a rectangular plastic crate when laser light penetrated its brain, at which point it began to run around in large circles. When the light was turned off, the mouse stopped running and started sniffing again. In other words, Zhang was able to control the movement of a living mouse by remote control.

'It sounds like a science-fiction version of stupid pet tricks,' the reporter later wrote.[74] But from this, as Zhang recalls, 'We knew that [optogenetics] could drive very, very robust behaviour.'[75] 'This was the moment,' Deisseroth said, ' … we finally knew that we had made something … broadly applicable.'[76]

Deisseroth is one of a rare group of neuroscientists who are also psychiatrists. His mission is not, of course, to turn animals into something to play with. He wants to tackle conditions such as depression, Alzheimer's, schizophrenia and autism.

Our most common mental health problems relate to anxiety. Of course, some level of anxiety is a normal human experience which can help us react well to emergencies or difficult situations. But anxiety can become excessive and pathological. Studies vary in estimating how prevalent this is, but something like a third of us are affected by an anxiety disorder during our lifetime, including panic attacks, phobias or obsessive-compulsive behaviour.[77] It may seem like anxiety disorders are becoming more common as a consequence of modern life, but there's no robust evidence for this. Large studies with the same methods haven't been repeated over any significant timespan to check this carefully.[78] Needless to say, anxiety disorders are not well understood. They are more common for women than men, but the reasons behind this aren't clear – and could include, for example, the additional workload women are more likely than men to have in caring for others.[79] Working out what actually happens in the brain to cause anxiety disorders is vital in all sorts of ways: for removing stigma, for helping us identify and treat problems, and for knowing what constitutes a healthy lifestyle.

Anxious people tend to show greater activity in a part of the brain called the amygdala, two almond-shaped clusters of neurons close to the hippocampus. This part of the brain is thought to be involved in emotional responses and dealing with threats. Animals which lack an amygdala don't exhibit normal fear responses. To try to better understand anxiety disorders, Kay Tye, a postdoctoral researcher in Deisseroth's team, used optogenetics to test what would happen if she manipulated the activity of neurons in the amygdala of mice.

Of course, it's impossible to know to what extent the feelings of human experience are also felt by animals, or to what extent our behaviours are comparable. Even so, scientists consider mice as being somewhat naturally 'anxious', in the sense that they like to hide away and avoid open spaces. A standard way of assessing just how anxious a mouse 'feels' is to use the so-called 'elevated plus maze' test. In this, mice are left to freely run in a cross-shaped track which is elevated above the lab bench. One axis has high walls and is relatively sheltered, while the other has low sides and is exposed. Mice tend to run up and down the protected axis of the plus shape, avoiding the open path. However, when Tye stimulated a particular set of neurons in their amygdala, mice suddenly became happy to explore the open environment.[80] Nothing else, such as their speed, seemed to change, just their sense of risk – or whatever it was that made them want to be in the more sheltered environment. In other words, Tye had discovered an 'anti-anxiety switch' that could be flicked on, resulting in the mouse becoming more explorative and, perhaps, feeling freer.

Deisseroth's team next developed what has become an important feature of many optogenetics experiments: rather than targeting neurons according to their location, scientists can now choose them according to where they reach. This is achieved by delivering the light-switchable protein to one area of the brain, and the light itself to somewhere else in the brain. This way, only neurons which span the two areas became activated. Using this, Deisseroth's team showed that each particular feature of anxiety – including increased respiration, risk-avoidance and apprehension in the absence of any actual risk – involves neurons connected to different brain regions.[81] Another team of researchers showed that the neurons involved in anxiety also link to parts of the brain involved in motivation.[82] In other words, separate brain modules are involved in different aspects of anxiety.

There's no escaping the vital caveat that all this work is done with mice, not humans, and carried out in unnatural settings. And what it tells us about the causes or nature of anxiety is very hard to say. But if any comfort could be drawn for anyone with an anxiety disorder, it's perhaps that these behaviours in mice could be dialled up or down instantaneously. It's not that optogenetics could easily do this in people, but that these discoveries tell us which brain neurons might be best targeted for investigation. Most importantly, they point to new therapeutic ideas – and to change being possible.

One area where this already looks especially promising is in treating addiction. In 2001, a team led by Antonello Bonci, then at the University of California San Francisco, found that mice given a single dose of cocaine showed a change in brain activity.[83] Circuitry normally involved in reinforcing learning was affected for several days. A broad implication was that a vulnerability to addiction might be opened up by just a single dose of cocaine. Then, in 2013, Bonci and their many collaborators showed that, at least in rats, optogenetics could be used to alter brain activity to stop cocaine addiction.[84]

For eight weeks, rats were given access to cocaine by pressing a lever. Then the set-up was changed so that each time a rat pressed the lever to obtain cocaine, there was a chance it would also receive a mild electric shock to its feet. This was enough to stop some rats from continuing to take cocaine. But others would self-administer the drug even when doing so had this clear negative consequence. In those rats, the team found a particularly low level of activity in their prefrontal cortex, a region of the brain implicated in all sorts of behaviours. Astonishingly, optogenetic stimulation of that region rescued the rats from their addiction.

Needless to say, human and rat brains are very different. Human brains are much larger, for example, and are especially well developed for language, while rat brains are more adept at dealing with smell. But perhaps surprisingly, there is a lot of commonality too. The broad set-up of the brain is similar and many basic pathways seem to be shared. Something as primal as the feeling of pleasure leading to wanting to repeat an activity again – the so-called reward pathway – is so essential for survival in us and other animals that at least something of its structure is preserved across species. This is the pathway hijacked by addiction. So it's at least feasible that what helps stop an addiction in rats could help people too.

Optogenetics can't be used directly on humans, not least because it would require genetic modification to a person's brain. But the activity of a human brain can be influenced instead by transcranial magnetic stimulation (TMS). Here, a small device that produces a rapidly changing magnetic field is placed against the head, inducing local electrical activity in the brain. The device is sometimes used to treat depression where drugs or psychotherapies have failed. In 2016, a group of cocaine addicts were treated with TMS, targeting the part of their brain analogous to the area in rats' which had been targeted by optogenetics.[85] The effect was clear: stimulating this part of the brain suppressed their urge to want cocaine.

As striking as this is, there isn't yet consensus as to whether or not this technique can play a role in medical care. In this study, patients knew they were being treated and so could have also benefited from a placebo effect. More work is also needed to standardise the procedure.[86] But even so, it's clear that discoveries made by optogenetics are important.[87] Probing the brain in this way is almost certainly going to help us better categorise mental health issues. And eventually, this seems likely to lead to new treatments that specifically target the right brain circuitry, replacing the current gamut of broad-brush pharmaceuticals.

Most people would be happy with optogenetics, and other technologies, leading us to new medicines to treat clear problems. But in-depth understanding of the brain is likely to also lead us to far more controversial issues. Intelligence is something of a taboo topic in contemporary science. It's hard, if not impossible, to define it, let alone measure it. Supposed tests of intelligence only report how good someone is at that type of test. Still, there is plenty of evidence to show that people are interested in boosting their own cognition. One in five respondents to a survey in the journal *Nature* in 2008 said that they had used drugs to stimulate focus or improve concentration.[88] People are already self-medicating to boost their cognition, despite so little clear information about what does or doesn't work, or even what the aim is.

It's hard to say whether or not we will understand what 'riding a bike' looks like in a brain anytime soon. But there can be no question that technology will continue to develop and, as our knowledge deepens, new ways of manipulating the brain are going to be more powerful and more precise.

The global endeavour to map the detailed structure of a brain is well under way, and already a mouse can be instructed to move by remote control. The aim of all this is to solve problems like depression or anxiety disorders. But mission creep is inevitable. For now, our vulnerabilities, our vanity and our capacity for love and hate, remain hidden inside our brains. There is a door which hasn't yet been found, a code still to be cracked and a threshold which hasn't yet been breached, but it's only a matter of time.

A conversation, book, song or movie – any number of things affect you. But for good and bad, hacks into the brain – in the form of virtual reality headsets, manipulative advertising and repetitive games on a touch screen – are becoming much more direct. All of which is nothing compared to what's coming. In twenty, fifty or a thousand years from now, our understanding of the human brain will be at a whole other level. We are at the moment before a jump scare. Something big is about to happen. We're fumbling in the dark, the tension is building, we know it's coming, but it's hard to say when – or what it will be.

5 The Others Within

My teeth are kept usually very clean, yet when I view them in a magnifying glass, I find growing between them a little white matter ... [and] to my great surprise perceived that the aforementioned matter contained very many small living animals.

Anthony Leeuwenhoek, letter to the Royal Society,
17 September 1683

In the 1970s, it was thought that about 300 different species of bacteria might be found in the human body. With this in mind, scientists at the time set out to identify a core set of bacteria found in healthy people, thinking that if any of these were missing, it would indicate disease, or perhaps even be an underlying cause of disease. Now we know that this idea was far too simplistic. In fact, the human body hosts an ecosystem of microorganisms of unimaginable diversity. There are about as many individual bacteria in you as there are human cells. They comprise around 10,000 different species of bacteria, some of which are not known to exist anywhere else on Earth. Altogether, these bacteria carry about 1,000 times as many genes as your own human genome. As well as this, there are untold numbers of viruses and fungi in, and on, our bodies, about which we know far less than we do the bacteria. In total, this – the human microbiome – amounts to something akin to an organ weighing about the same as a human brain. The reason it's not easy to relate the contents of a person's microbiome to disease is that this vast universe of life within our bodies is enormously diverse and – unlike any other human organ – it varies considerably from person to person, and changes during the course of our lives, as we go through puberty, pregnancy or even when we move house.

In the last decade or so, our understanding of the microbiome has exploded, thanks to the development of two types of technology which together identify microbes genetically. First, laboratory hardware can be used to rapidly sequence large amounts of genetic material.[1] Secondly, we have developed computer hardware and software that allow us to sort out all the different microbial gene sequences, seek patterns in the data and correlate results with other factors, such as a person's diet or state of health. The endeavour to understand the human microbiome has become a flagship enterprise for big-data science.

Even though the full extent of the symbiosis of humans and microbes is yet to be realised, there is no doubt that our health and wellness vitally depend on their alliance. And like any long-term relationship, the bond is complicated. Different parts of the body are colonised with different microbes, for example. Those that live on our teeth are different from those on our skin or in our gut. Even in a person's gut, there is exquisite diversity in the types of bacteria that live along its length. Separate environments, like islands hosting different animals, are created by folds in the intestinal wall and local variations in acidity, mucus and oxygen.

It is these, our gut bacteria, which have been studied the most, mainly by analysing faeces.[2] As long ago as the 1680s, when the Dutch scientist Leeuwenhoek first used a primitive microscope to discover bacteria, he looked at them in his own faeces. This must have been a shocking observation; intuition does not tell us that 25–50 per cent of our stool is comprised of living and dead bacteria. In 1909, the US bacteriologist Arthur Kendall suggested that the types of microbes in a person's gut could vary in accordance with their diet.[3] He had been testing the idea by feeding monkeys different foods and then trying to culture the bacteria in their faeces. From the decades of research that followed, the gist of what gut bacteria do for us – or at least one aspect of the function they provide – is now widely known: the gut provides a home for bacteria in return for their help in digesting food and producing nutrients. For example, gut bacteria produce B vitamins which we otherwise might lack.[4] But recently, our advancing knowledge has thrown up other revelatory details, far beyond anything Kendall might have imagined. From them, various claims have been made for microbiome-based therapies or treatments. Many are over-hyped, but not all; genuinely transformative ideas are on the horizon relating to everything from nutrition and diet to our ability to fight disease and even our mental health.

You might think that the shape of your body is a product of your genes, the food you eat and the frequency with which you exercise. But evidence has accumulated which shows that something else is also a major factor: gut bacteria. The journey towards this scientific revelation began in 2004 at Washington University, where a postdoc researcher in the lab of biologist Jeffrey Gordon observed something unexpected and important.

While working for his PhD in Sweden, Fredrik Bäckhed had become fascinated by the fact that the exact same bacteria which live happily in our gut can cause disease if they infect another part of the body, such as the urinary tract. He reasoned that there had to be something different about the way the gut senses and controls bacteria compared to other places in the body. He knew Gordon by reputation, and knew he was interested in studying gut microbes, so emailed him to ask if he could join his lab once his PhD was complete.[5] Bäckhed was full of ideas for research – such as looking at the effects of microbes on the nervous system – but together they settled on a plan to look at how microbes could affect an animal's metabolism. Gordon suggested they look at something very simple at first: whether the amount of body fat an animal has might be affected by the absence of *any* bacteria in its gut.

Normally, all mice have plenty of bacteria in their gut, just as we do. But in labs, mice can be kept in a sterile environment so that they never contact microbes at all. These mice are born and raised in a sealed-off plastic enclosure and fed food which has been irradiated. They have an unusually straightforward scientific name: 'germ-free mice'. What Bäckhed observed in 2004 was that germ-free mice were much thinner than microbe-laden mice raised in a normal lab environment. More importantly, when he deliberately exposed germ-free mice to bacteria, they began to put on weight.[6] It's not that they started eating more – in fact, they ate a little less – they just got fatter.

On the face of it, these results implied that microbes directly affect an animal's body weight. But as germ-free mice are not found naturally, any number of strange things might have been going on in these animals, with their fluctuations in weight being mere side-effects. Evidence of a different kind was needed to establish a direct causal link, and this came from another postdoctoral researcher in Gordon's team, Ruth Ley. In her experiment, instead of germ-free mice she used mice with a specific genetic mutation that causes obesity. Specifically, these mice had a non-working version of the gene responsible for the production of a hormone called leptin, which helps the animal match its energy intake to its energy use. A deficiency in this hormone causes the mouse to uptake more energy from its food than it needs. By studying the microbiome of mice with this genetic mutation and comparing it to those of mice without it, Ley found that the obese mice harboured a distinct mix of bacteria in their gut.[7] Even among littermates given the exact-same food and living in the exact-same environment, those mice that inherited this genetic mutation had a different microbiome from their siblings.

It could be that the changes in their microbiome and in their body weight were both caused by this genetic mutation inde-pendently, but it seemed more likely that the two were connected: that it was the mutation that caused mice to become obese, which in turn changed their microbiome, or that the mutation affected the microbiome, which then caused the mice to gain weight. The next experiments in Gordon's lab gave the answer.

Working with Ley and others in Gordon's lab, PhD student Peter Turnbaugh transferred bacteria from overweight or lean mice into germ-free mice.[8] Amazingly, mice which received microbes from overweight mice put on far more weight than those given microbes from lean mice. Two weeks after being colonised with bacteria from overweight mice, their body fat increased by an average of 47 per cent. Those given microbes from lean animals also increased their body fat, but to a much lesser extent. These results implied that gut microbes could directly influence the size of mice. This happened more or less irrespective of how much food they ate. If anything, mice given the microbiome of an obese mouse ate slightly less. As Gordon later recalled, 'It was an 'Oh my God' moment.'[9]

By analysing the microbes in detail, the team found that a high proportion of bacteria from obese mice included enzymes that break down sugars that are otherwise indigestible. This suggested that microbes from obese mice equip the animal with an ability to harvest more energy from food. Amazingly, these same types of bacteria – those rich in enzymes which break down sugars – were then also found to be more abundant in obese people.[10] Later, leading her own lab, Ley showed that microbes from obese humans also cause germ-free mice to gain far more weight than if they received microbes from lean people.[11] Altogether, this led to a revolutionary idea: that the composition of a person's gut microbes can affect how much energy is extracted from food, which can in turn impact a person's body weight. This ground-breaking discovery hinted at new possibilities for healthcare, but a further step would be needed before they could be realised.

Eran Elinav, based at the Weizmann Institute of Science in Israel, sees it as his life's mission to understand the molecular language by which the human body talks with microbes.[12] He trained to be a medical doctor, his childhood dream, but 'clinical routine got to be a little bit boring'.[13] So he switched to focus his career on research, and earned a PhD in immunology. He became interested in the microbiome in the mid-2000s, just as research in the subject was taking off. Leading his own lab since 2012, he set out to understand the relationship between diet, obesity and the microbiome.

As Elinav learnt from countless scientific papers about nutrition, many diets originate in a system for rating foods according to their effect on our blood sugar level: the glycaemic index. This way of characterising food came from research led by David Jenkins at the University of Toronto in 1981. Jenkins had small groups of people fast overnight, eat a particular food, and then measured their blood sugar levels over the following two hours.[14] Each type of food was given a score according to how much it raised those levels per unit of carboydrate, with sugar as the benchmark with a score of 100. Honey scored 87, sweetcorn scored 59, tomato soup 38, and so on. The reason this was an important advance was that it rated food not by what it's made of, as such, but by how it affects the human body. Today, every conceivable edible thing has been analysed this way and – very generally speaking – those seeking to lose weight are advised to avoid foods with a high glycaemic index, which cause short-lived spikes in energy that may soon leave us craving more, and to tend toward foods with a low glycaemic index, which release their energy more slowly, helping us feel fuller for longer.

The reality is a lot more nuanced, of course, as it always is. According to the original 1981 analysis, for example, carrots score far higher than white bread on the glycaemic index because a certain amount of carrot carbohydrate causes a far higher rise in blood sugar levels than the same amount of white bread carbohydrate, but a huge number of carrots would have to be eaten to provide the same amount of carbohydrate as a piece of bread. Another score – called the glycaemic load – has been invented to take this into account, by multiplying the glycaemic index of food by its total carbohydrate content. This gives carrots a score much lower than that of white bread. But neither the glycaemic index nor the glycaemic load take into account, for example, a food's vitamin and mineral content. Despite these and other caveats, countless diet plans have built on this way of ranking food. In fact, as Tim Spector, author of *The Diet Myth*, has put it: 'The ritual of dieting has become an epidemic.'[15] And yet every decade, across the globe, humans are getting heavier.[16]

We must remember that in many parts of the world, increasing weight is good news: the under-nourishment of children in many countries, such as Brazil and Bangladesh, has fallen considerably in recent years.[17] Still, a shocking number of children die each year from causes which are preventable with adequate nutrition. This is all the more heartbreaking when the amount of food being produced in the world has risen so dramatically. At the same time, the number of children across the globe who are obese has increased so much that it is now similar to the number who are under-nourished.[18] And of course, the weight-loss industry in many countries is enormous.

Part of the reason for this is that no single diet plan has ever proved to be better than the rest. What works for one person doesn't necessarily work for everyone else. And we've all come across someone who seems to stay a healthy weight no matter how much cake, chocolate or wine they consume. This variation holds true in carefully controlled research too. In one clinical trial involving over 600 people, two different diets – low-fat versus low-carb – were pitched against each other over a twelve-month period. It turned out that, on average, both diet plans had a similar effect, even while each person's individual response varied hugely; some gained and others lost weight, some a little, some a lot.[19] Understanding this – the differences between us – is where Elinav made a vital discovery.

He didn't get there alone but with Eran Segal, also at the Weizmann Institute. They were introduced to each other by a mutual colleague: 'Trust me, this is a great guy who has developed interests very close to yours.'[20] Segal had been studying genetics, but also had a keen interest in nutrition, not least because he was a marathon runner married to a dietician. Just like Elinav, he had read a lot on the subject and was well aware of the competing claims and counterclaims about the merits of one or other diet. 'What better way to sort it all out than with big data and a computer algorithm?' he thought.[21] Together,

Elinav and Segal spent around a year talking, getting to know one other and familiarising themselves with the other's specialist jargon.[22] Their different perspectives – computer science for Segal and biology for Elinav – were crucial to their success. Together they decided that if they could collect and analyse enough information about enough people, something big had to unravel.

Their broader aim was to find out what kind of diet might help a person lower their risk of developing a problem like type 2 diabetes, which is characterised by heightened blood sugar levels. (There is evidence that avoiding foods which cause blood sugar levels to spike can help reduce this risk.[23]) The more immediate task was to undertake an in-depth, large-scale study of how people's blood glucose levels are affected by eating. At first, they took measurements of themselves and a few others to check that they could do this kind of work.[24] Then they embarked on doing the same on an unprecedented scale.[25] Instead of studying a handful of adults, as was done in 1981, Elinav and Segal recruited over 800 people. They didn't pay people to participate, but many readily signed up because they wanted to learn about themselves. And instead of taking glucose measurements a few times over the course of two hours, every person's blood sugar level was measured every five minutes over seven days. Altogether, this level of scrutiny would amount to more than 1.5 million glucose measurements.

A small sensor developed for monitoring glucose levels in people with diabetes was attached to everyone involved, usually on their abdomen. This type of sensor uses a very thin needle, about the size of an eyelash, to reach into the liquid just under the top layer of skin, where glucose levels mirror that of blood. In fact, this liquid – called the skin interstitial fluid – is a rich source of molecules that can indicate a person's state of health; in the future, this type of sensor might be used for all kinds of diagnoses.

Participants carried on with their everyday lives but were told to log all their activities via a phone app, including when they slept or exercised, as well as what and when they ate. The only exception to their normal routine was that each morning they were to eat a prescribed breakfast that consisted either of bread alone, or of bread and butter, or of 50g of sugar taken as either glucose or fructose. This gave the team data on people's responses to 46,898 regular meals and 5,107 set breakfasts. As well as this, each participant answered a detailed medical questionnaire, was subject to a variety of physical assessments such as measurements of their height and hip circumference, and all of them had their stool analysed for the make-up of their microbiome.

It turned out that glucose levels spiked exactly in accordance with each food's glycaemic index. But crucially, this was only the case *on average*. The variation from one person to the next was enormous. For any given food, some people's glucose level would spike dramatically while others hardly seemed to react at all. This couldn't be explained away as a random fluctuation, because the same person responded similarly each time they ate that particular food. For one middle-aged woman, for example, her blood glucose level spiked every time she ate tomatoes – a food that she had, in fact, been eating lots of as part of a specific dietary plan.[26] Another person spiked especially strongly after eating bananas. 'We had stumbled across a shocking realization,' Elinav and Segal later wrote: 'everything was personal.'[27]

Segal's wife, Keren, was especially stunned. As a dietician, she had been trained to provide guidance to countless people about what they should and shouldn't eat. Now, her husband had evidence that her dietary advice might not have always been helpful. That some people's post-eating sugar levels spiked more in response to rice than ice cream was especially striking to her. It dawned on her that she might have even directed some of her patients to a type of food which, though beneficial on average, was wrong for them personally.[28]

With all the information they had gathered, the team set out to see if people's post-meal glucose levels correlated with anything else, such as their weight, size, age or the amount of sleep or exercise they got. A machine-learning algorithm was used to ascertain which combination of factors needed to be taken into account to generate the most accurate forecast of a person's post-meal glucose response. Digging into what the computer came up with, it turned out that a complex combination of factors, including a person's age and body mass index, were involved, but one factor stood out as the most significant contributor by far: each person's microbiome.[29]

Having discovered the crucial factors involved, Elinav and Segal decided to test whether it was possible to design a tailored plan that successfully controls an individual's blood sugar levels. They took on a group of twenty-six pre-diabetic volunteers – people whose glucose levels had a tendency to fluctuate significantly but who weren't diabetic – and carried out all the same analysis as before. Various physical and blood measurements were taken, their microbiome was scrutinised and their glucose levels were monitored over the course of a week. Then, all this personal information having been analysed, each of them was given two personalised diet plans, each to be followed for a week: one 'good' diet plan that was designed to keep blood sugar levels low and relatively stable, the other a 'bad' diet plan that that would in theory lead to their glucose levels spiking up and down. The participants didn't know which was which, and it wouldn't have been obvious. One person's good diet included, for example, hummus, pitta bread, eggs, noodles and ice cream, while their bad diet included breakfast cereal, sushi, sweetcorn and chocolate.[30] On account of everyone being different, no two people were given the same diet plan. It was even the case that a type of food in one person's good diet would be in another person's bad diet.[31] What happened was exactly as Elinav and Segal predicted: when people were on their good diet week, their glucose levels were low and stable, while on their bad diet week, they fluctuated much more.[32]

Crucially for our purposes, they discovered that people's microbiome changed between their good and bad diet weeks, and some of those changes were similar, even when the specific foods being eaten were different. Three types of bacteria which increased in a number of people while on their good diet plan turned out to be bacteria that people with type 2 diabetes tend to have less of. This was consistent with the idea that a good diet favours a microbiome composition which correlates with protection against type 2 diabetes.[33] Within one day of these results being published, over a hundred articles were published discussing its implications.[34]

For decades, Western societies have been trying to tackle obesity and obesity-associated diseases like type 2 diabetes with diet plans. But Elinav and Segal's research suggests a huge problem. What constitutes a 'healthy' diet depends not only on what food is being eaten but on who is eating it: their genetics, their lifestyle and, perhaps especially importantly, their microbiome. We need to move away from a 'one size fits all' approach. 'We're entering a new era of nutrition,' Segal says, 'of, "What is the best diet for me?"'[35]

In December 2017, Elinav and Segal published a book, *The Personalized Diet*, in which they envisage a scenario in which everyone scrutinises their own blood sugar levels after eating different foods. This can be done using a finger-prick test, the type readily available over the counter for diabetics.[36] By doing this, each person could devise their own healthy diet plan to avoid foods which make their own blood sugar levels spike. In the longer term, however, they hope it will be possible to do this another way: for people to answer a questionnaire and mail off a stool sample, so that a computer algorithm can use their answers and the contents of their microbiome to predict a personalised healthy diet plan in return. Indeed, the computer algorithm they developed for their experimental work has been licensed to a company with just this aim.[37]

In the meantime, Elinav and Segal's work doesn't lead to any simple takeaway advice about what to eat or avoid. Their research doesn't even conclude that one particular type of bacteria is vitally important. Rather, it points to certain trends in the types of bacteria present in a person's microbiome, and perhaps especially its overall diversity, as being healthy. More importantly, neither does this research tell us anything directly about the risk of developing type 2 diabetes. Testing the effects of diet on hard health outcomes is notoriously difficult, and a robust test of this would require the monitoring of the health of large numbers of people, each sticking to a particular diet, over many years.[38] For this to be done with the same kind of rigour used to test a new medicine – in a double-blind clinical trial – the participants wouldn't even be able to know what they were eating, which is obviously impossible. This is why almost all trials of dietary interventions are short-term and have to use blood tests as a proxy for actual disease risk.[39] Indeed, while there is evidence that avoiding foods which cause blood sugar levels to spike can help reduce the risk of developing type 2 diabetes, this isn't proven beyond doubt.[40] With such difficulties, simple health messages are hard.

What does emerge clearly is a picture in which the microbiome, diet and physiology of the gut are deeply entwined. Each is exquisitely complex on its own terms; how they interact even more so.[41] Everyone's microbiome involves countless different types of bacteria, including some we haven't yet identified.[42] Their diet involves thousands of chemicals eaten in varying amounts at different times of day. And every person's basic physiology is shaped by their genetics, the state of their immune system, their history of infections and more. For science, and for our understanding of what constitutes healthy eating, it is this complexity, which can only be grasped and analysed with the help of big data and computer algorithms, that now sets the direction of travel. Elinav and Segal's research shows how the science of diet, nutrition and the microbiome is entering a revolutionary phase.

As it does, we will be forced to reckon with a host of political, social and ethical questions. As anyone who has ever tried any kind of diet knows all too well, our eating habits are not solely driven by *knowing* the types of food which are good for us. Global corporations thrive on producing and selling foods and drinks with very tasty combinations of fat, salt and sugar, which can seem irresistible.[43] Government policies can help counterbalance commercial interests as they have done with smoking.[44] Recently, several governments have tried to lower people's sugar intake by introducing a tax or levy on high-sugar drinks. Hungary was first in 2011, and France followed in 2012. The UK introduced this in April 2018, and soon several companies reformulated their drinks to contain sugar at a level below the threshold for being taxed.[45] At least in the UK, where most healthcare is paid by taxation, there is a simple financial argument for doing so: in the UK, the annual cost of direct care for people with type 2 diabetes is estimated as £8.8 billion. But if Elinav and Segal's vision is realised and it becomes clear that personalised nutrition would have a huge impact on human health, the question will present itself: should analysis of a person's microbiome and a personalised diet plan become part of routine, preventative healthcare, perhaps paid for by taxation? Where do we draw a line between a nutritional product, a dietary plan and a medicine? As nutritional interventions and diet plans become more advanced, they must surely be tested with the same rigour we use for regular pharmaceutical medicines, and then deployed with fairness. As any science matures, new policies must be developed. This will be especially important when it concerns such a vital part of our daily lives: what we eat and drink.

Needless to say, the microbiome is important to us beyond the realm of diet and nutrition. In fact, there's scarcely any state of human health or disease that hasn't been linked with it. Variations in the human microbiome have been associated with diseases as diverse as autism, asthma, multiple sclerosis, cancer and inflammatory bowel disease.[46] Importantly, however, these are (so far) only correlations. It is very difficult to test whether or not variation in a person's microbiome *directly* causes disease or worsens symptoms.[47]

Many labs have tried to do this by studying the effects of transferring human-derived microbes into germ-free mice. One experiment, for example, used microbes extracted from the faeces of thirty-four pairs of identical twins, where one twin had multiple sclerosis and the other didn't. After being washed, the microbes were transplanted into mice already predisposed to develop a disease with symptoms like those of multiple sclerosis.

Strikingly, mice receiving bacteria from a multiple sclerosis-affected twin were much more likely to develop the disease themselves.[48] Along the same lines, microbes from people with inflammatory bowel disease were transplanted into mice susceptible to the same illness. This time, the transferred bacteria caused the animal's symptoms to worsen.[49] These results, and others like them, suggest that the microbiome can directly affect the likelihood of disease developing or the severity of symptoms. But how this one factor could affect so many different types of illness isn't obvious.

That said, there is one aspect of the human body already known to affect how we fare in all sorts of diseases: our immune system. So one way in which the microbiome could affect us in so many ways is by having an effect on the immune system. One of the first to suggest a general link between our exposure to germs and the state of our immune system, which is potentially of huge importance, was the epidemiologist David Strachan of St George's Hospital, London. It is known as the hygiene hypothesis.

Scrutinising a survey of over 17,000 children, Strachan noticed that whether or not they developed hay fever correlated with the size of their family, and especially the number of older siblings they had.[50] The bigger the family, the less likely it was that they would develop the allergy. He theorised that larger households are likely to be hit by more infections, and perhaps this increased exposure to infection during early childhood might help protect against hay fever in some way. This led him to the bold idea that, in general, something about a 'dirty' environment early in life could help prevent allergies. The hygiene hypothesis still guides our thinking today, but our view of it has changed because of something Strachan couldn't possibly have foreseen: the importance of gut microbes. It was once thought that immune cells never came into direct physical contact with bacteria in the gut, or if they did they simply ignored (rather than attacked) them. We now know this isn't true.

Microbes are present all along the human gut – the oesophagus, stomach, small intestine, large intestine and rectum

– although their abundance varies a lot. Many types of bacteria are killed, or held back from multiplying much, by acid in the stomach, so there are relatively few living there. After the stomach follows the small intestine, the 6-metre-long folded tube where most nutrients are absorbed, and some bacteria live there. But by far the greatest numbers of bacteria reside in the final stretch of food's journey, the 1½ metres of gut called the large intestine.

The inside of the large intestine is lined with cells, called epithelial cells, coated with a layer of thick mucus. Some bacteria penetrate this thick mucus, but most live within a more fluid layer of mucus which lies on top, or on the mucus surface. Immune cells, on the other hand, are located in the tissue underneath the epithelial cells. They are not inside the intestinal tube where the bacteria reside, but in the tissue surrounding it. There, they are poised to protect us against any bacteria which try to breach the layer of epithelial cells and invade the body. This set-up seems to imply that immune cells won't come into direct contact with gut bacteria unless the bacteria attempt to leave the gut. In fact, as we now know, immune cells have protrusions which pierce through the layer of epithelial cells lining the intestine and directly contact bacteria living in and on the mucus.

The question is: why don't immune cells react against these gut bacteria as they would against bacteria elsewhere in the body? It's not that the immune cells are of an entirely different type to those found elsewhere in the body, or that the bacteria are. Rather, it's that something in the environment of the gut causes immune cells to behave differently. Crucially, when immune cells detect bacteria in the gut, they not only refrain from attack, but also actually secrete chemicals and proteins that serve to maintain gut health.[51]

This requires us to look at the immune system in a different way. We tend to think that the immune system's mission is to destroy disease-causing bacteria, viruses, fungi and other intruders. But while it does indeed do this, that's not all it does. We've already seen that in the womb, for example, the immune system helps build the placenta during pregnancy. Likewise, the immune system takes on other jobs in the gut, including maintaining the epithelial cell lining of the large intestine and controlling the types of bacteria allowed there.

In turn, gut microbes help develop and sustain our immune system. One way they do this is through the production of molecules called short-chain fatty acids. These molecules are produced in the chemical reaction by which bacteria gain energy from breaking down plant fibre.[52] In detail, gut bacteria produce high levels of three types of short-chain fatty acids: acetic acid, propionic acid and butyric acid or butyrate. The last of these, butyrate, promotes the activity of immune cells called regulatory T cells, or T regs (said like the dinosaur T. Rex).[53] These cells are specialists at turning *off* the activity of other immune cells, a vital action in order that the immune system doesn't damage the body. The other short-chain fatty acids also affect immune cells, as well as the epithelial cells of the gut lining, although these other processes are less well understood. Overall, roughly speaking, high levels of these three types of fatty acid molecules quieten down the immune system, creating an 'anti-inflammatory' environment. What is particularly amazing is that these molecules seem to affect not only the local immune cells of the gut but the whole body's immune system.

Allergies are caused by undesirable immune responses against things mistakenly seen as harmful when they aren't really – what we might think of as over-reactions of the immune system – so something that helps dampen immune responses, or helps the body develop the capacity to do so, would in theory be helpful in preventing allergies. Clearly something about the composition of the gut microbiome helps the immune system develop in precisely this way. In support of this, mice given a high-fibre diet produced high levels of short-chain fatty acids in their gut which correlated with them being less likely to develop a mouse version of asthma.[54]

One plausible process, consistent with what's known so far, is that a high-fibre diet boosts the numbers of gut bacteria which thrive on fibre, which leads to high levels of short-chain fatty acids being produced. As well as acting in the gut, these molecules circulate through the whole body in blood, allowing them to affect immune cells in different organs. There is some evidence, however, that they mostly affect immune cells in bone marrow – potentially another clue, as the bone marrow is an important place where immune cells develop. Perhaps their exposure to short-chain fatty acids there helps to set their re-activity at the right level before they travel out from bone marrow into the body's tissues and organs. Without this happening, the unwanted immune responses that underlie aller-gies are perhaps more likely. Consistent with this, a small study of young children found that those with allergies had lower levels of short-chain fatty acids in their faeces.[55] But again, this whole process is only plausible, not proven. This observation in children, for example, is only a correlation.

Even if the microbiome is taken to be important in the devel-opment of the immune system and the likelihood of developing allergies, we don't yet know how big an effect it has in compari-son with everything else that also affects our disposition towards allergies – genes, smoking, age, exposure to allergens and much else. We are at the frontier of knowledge here. We have scraps of information, not yet sufficient to develop microbiome-based medicines for allergies.

There is, however, one type of medicine based on the micro-biome which has been proven to work, not for allergies but to stop an infection – faecal transplantation.

A faecal transplant is as simple and as strange as it sounds. The process varies depending on where it's done but, roughly speaking, works as follows. A fresh stool sample is collected and passed to a lab. It's then whizzed up in nothing more high-tech than a household blender. After being sieved – to be sure any lumps are removed and it's a nice smooth consistency – it's sucked up into wide plastic syringes. A blood test of the donor needs to have been performed in order to check they don't have hepatitis, HIV or other infections, and the stool itself is tested for infections and parasites. Assuming it gets the all-clear, it's time for the transfer.

An intravenous drip delivers sedatives to the patient so that they are asleep or nearly asleep. They will have had a liquids-only diet for many hours and perhaps a laxative the night before. A long flexible tube – a colonoscope – about one centimetre in diameter is inserted through their rectum. A camera transmits a picture from the end of the tube so that it can be positioned at the top of the large intestine. Once in place, a faeces-loaded syringe attached to the other end of the tube is pressed and the material slowly delivered. Sometimes, the sample is from a partner or family member, sometimes it's from a stranger.

Variations in this procedure include packaging the faeces – or sometimes, bacteria that have been isolated from faeces – into capsules which can then be delivered rectally or even swallowed. Another option is to deliver the transplant directly into the patient's stomach via a tube fed through their nose. In this case, a drug is also taken to stop the stomach from producing acid which would otherwise kill much of the transplant. Needless to say, whichever method is used, the process can be embarrassing and worrying. Perhaps this is one reason why faecal transplants are sometimes self-administered at home, even though it's far safer to involve healthcare professionals. Do-it-yourself instructions are readily available online, but are not recommended, for reasons we'll soon see.

Amazingly, this relatively low-tech procedure is beneficial in at least one life-threatening situation: recurrent gut infection with a bacterium called *Clostridioides difficile*, commonly known as *C. diff.* First isolated in 1935, *C. diff.* was named for being a type of bacterium which was *difficult* to isolate and culture.[56] Normally, a *C. diff.* infection can be treated with antibiotics, but some strains have become resistant. The resistant strains of *C. diff.* are 'super-bugs', and they are causing problems with increasing frequency.[57] Common symptoms include cramps, fever and severe diarrhoea, and most people recover, but if a *C. diff.* infection isn't brought under control, it can be fatal.

Surprisingly, *C. diff.* is naturally present in some people's gut. So it's not that this type of bacteria is inevitably dangerous. Problems are more likely to occur in elderly patients, people with certain other illnesses such as cancer or inflammatory bowel disease, or anyone with a weakened immune system, as can happen as a side-effect of being on chemotherapy or taking steroids. Paradoxically, broad-spectrum antibiotics can also increase the risk of *C. diff.* causing problems. This isn't a reason to avoid antibiotics; they are indispensable in saving lives by treating dangerous bacterial infections. But as an unwanted side-effect, they can attack bacteria other than those for which they were prescribed, including those normally resident in the gut. This loss of microbial diversity can give *C. diff.* bacteria – whether recently ingested or already present – an opportunity to flourish.

As an anecdote – not to be taken as robust or proven medical advice – one scientist studying this told me that when she had to take antibiotics for a long period of time, she adopted a diet especially high in fibre, to try and stabilise her gut microbiome.[58] Her thinking was that the high levels of fibre might encourage her normal gut bacteria to multiply, which, in principle at least, may help counterbalance these unwanted side-effects of antibiotics.[59] Whether or not this helps is anybody's guess; the only microbiome-based intervention widely agreed to work in the treatment of *C. diff.* infection is a faecal transplant of a 'healthy' microbiome directly into the patient's intestine. This is thought to work for two reasons.[60] First, the donated bacteria compete with *C. diff.* bacteria for nutrients and other resources. Secondly, there is an effect on the gut immune system, which in turn helps constrain *C. diff.*

While this understanding is new, the idea of a faecal transplant itself is ancient. As long ago as the fourth century, the Chinese medical doctor Ge Hong used stool from a healthy person to treat severe diarrhoea.[61] More recently, in 1958 – still long before much was known about the microbiome – four patients were treated with a faecal transplant for what was likely to be a *C. diff.* infection.[62] Ben Eiseman, Chief of Surgery at the Denver Veterans Affairs Medical Center, carried out the process because 'those were days when if one had an idea, we simply tried'.[63] His idea didn't take off because at that time antibiotics were still effective.[64] Now, with the rise of antibiotic-resistant *C. diff.*, the rationale for the procedure is much clearer. In 2013, one small trial in the Netherlands was so successful that it was stopped early, because it would have been unethical to prevent a control group of infected patients from receiving a faecal transplant purely for the sake of comparison with those who had.[65] Since then, many other trials have been similarly successful.[66]

Nonetheless, it is still something of an experimental procedure, and in 2019 one seventy-three-year-old man died as a result of a faecal transplant. His death was caused by the faeces he received being contaminated with a strain of *E. coli* bacteria that hadn't been screened for.[67] The bacteria were of a type that commonly causes travellers' diarrhoea, but the effects were more serious because the patient had a weakened immune system. Another patient was adversely affected by faeces from the same donor but, having developed a fever and cough, was admitted to hospital, given antibiotics and recovered. There are almost certainly other instances of infections being transmitted by faecal transplant that have gone unrecorded.

In the case of the 10,000 faecal transplants that are performed in US medical centres annually to treat recurrent *C. diff.* infections, the benefits outweigh these risks. But this is less clearly the case in the treatment of other diseases. Hundreds of clinical trials are under way to test whether or not faecal transplantation can help treat other infections, autoimmune diseases, psychiatric conditions and more.[68] If so – and it seems likely that it could help in at least some of these other situations – then one problem we will need to tackle is how to ensure a consistent and standardised treatment.

Faecal transplantation involves so many more variables than the taking of a pharmaceutical drug. As the name suggests, it is a lot more like moving an organ from donor to patient. Every microbiome is unique, and currently there are, as we have seen, a host of different methods available. It will be important to test their relative safety and efficacy. It may be preferable for clinics and hospitals to obtain faeces donations from specialist centres rather than collect their own. Already, there are a few faeces banks around the world, including the non-profit organisation OpenBiome, in Cambridge, Massachusetts, and the Netherlands Donor Faeces Bank.[69] They hope to do for faecal transplants what has been done by blood banks for blood transfusion.

Eventually, some kind of next-generation microbiome medicine is likely to be a better option. But there's a deep scientific problem here: that with such variability in the human microbiome, we don't really know what a 'healthy' one is. A core set of various bacteria seem important and, by definition, there must be an absence of anything obviously dangerous, such as an abundance of *C. diff.*[70] But beyond this, little is clear. Rather than a few particular types of microbes being needed, perhaps what's important is an overarching ecology or stability. If we understood this clearly then a healthy microbe cocktail could be manufactured by design. This would circumvent the variability and risks of using someone's faeces directly.[71]

Probiotics – foods or supplements with live bacteria added – are an alternative prospect for manipulating the microbiome, and there is some evidence they may ease the symptoms of an ongoing illness such as irritable bowel syndrome, or perhaps help avoid the side-effects of taking antibiotics.[72] But the relevant authorities across Europe and the USA have yet to approve any probiotic as a medicine. Most probiotics are sold as dietary supplements, which means they don't get tested in the same way as pharmaceutical drugs. Eran Elinav, as well as others, argues that this is wrong, and that probiotics should be tested in rigorous clinical trials so that we can establish whether or not they work.[73] They have potential, he says. But for now, there is no definitive proof that any probiotic food or supplement can manipulate a person's microbiome and treat a disease.[74]

There are grander radical ideas out there too, such as creating homes and offices which circulate microbes in the air, or in the water. Or spa pools which could contain a healing mix of bacteria. This is at the border between science and science fiction.

And while we're here, let's consider whether or not micro-biome-based interventions could go beyond treating infections – to hack the human brain.

Microscopic agents can definitely shift a person's behaviour. Rabies is a well-known example. It's caused by a virus that affects humans and dogs and has just five genes (compared to around 19,000 for a dog and 21,000 for a human). The virus produces proteins which interact with the nervous system, making an infected host become agitated and aggressive. An angry dog is then more likely to bite another dog, or a person, thus passing the virus on. In fact, there are all sorts of other examples of germs manipulating their host's behaviour to their advantage.[75] Some gut bacteria, for example, can induce a preference in flies for feeding on more of their fellow bacteria.[76] Other bacteria can affect a fly's desire to eat yeast.[77] These observations aren't well understood, but are likely to involve gut bacteria being able to affect the fly's nervous system, and perhaps the fly's sense of smell.[78]

Whether or not gut bacteria can deliberately influence human behaviour is a controversial topic. One study, involving just over 1,000 people, identified some types of microbe present in humans that we believe are associated with a high quality of life, and others that correlated with depression.[79] But again, correlation is not causation. Those conducting the experiment ensured that these effects were not due to some of their partici-pants taking antidepressants, but there are many other possible reasons for the pattern they detected. Some people with mental health difficulties may sleep less or eat irregularly, for example, which could affect their microbiome. Nonetheless, it is possible that gut bacteria affect our mental well-being directly, and they do produce neurotransmitters, like serotonin and dopamine. These molecules may reach the brain directly, but could also act locally, perhaps on the vagus nerve which connects the gut to the brain.[80]

One subtle theory argues that if our state of mind and behaviour are affected by the gut microbiome, it can only be indirectly, because gut microbes are unlikely to have evolved to control the human mind specifically. If we think of the gut microbiome as an ecosystem in which each type of bacteria is in competition with all the others for resources and living space, then it's hard to conceive of a way in which one type of bacteria could affect human behaviour to benefit only itself, for the reason that there are just so many types of bacteria in any one person's gut. If, for example, one type of gut bacteria made us prefer the food they thrive on, it is inevitable that others would also benefit from that food, undermining any competitive advantage.[81]

If there are indirect effects of the gut microbiome on our mental state, a likely candidate for an intermediary is the immune system, because everything in the body is so entwined with it. Indeed, we know for sure that activity of the immune system can trigger feelings of melancholy; this happens whenever you feel sick or have a fever. While much is unclear, one thing to emerge is that there is almost certainly no 'one size fits all' approach to treating depression, anxiety or any mental health problem. Perhaps in the future, a microbiome-based intervention will help in some cases. Already there's a new word for supplements containing living bacteria purported to have mental health benefits: psychobiotics.[82]

The skin, lung and mouth all have microbiomes too, each of which we know far less about than the gut's, and each of which contains another universe. And in this chapter we have only considered bacteria, overlooking all the other types of micro-organism such as fungi and viruses – not to mention types of viruses called phages that infect our resident bacteria – which, again, we know so little about. There may be life on other planets, but we know for certain there are aliens inside our own bodies. And arguably, they are of far greater importance. The knowledge we're building up about them is going to have an enormous impact on our lives, not tomorrow, and perhaps not the next day, but for sure in the twenty-second century.

6 Overarching Codes

New directions in science are launched by new tools much more often than by new concepts. The effect of a concept-driven revolution is to explain old things in new ways. The effect of a tool-driven revolution is to discover new things that have to be explained.

Freeman J. Dyson, *Imagined Worlds*

In the mid-1960s, a young professor at Caltech, William Dreyer, gave his first graduate student two pieces of advice. He told him to always practise biology at the cutting edge. And he said that if you really want to transform an area of science, invent a new technology.[1]

The student, Leroy 'Lee' Hood, took this guidance to heart. He had already begun to derive a deep satisfaction from pursuing new ideas, and his ambitions were growing.[2] Dreyer felt that his student was 'kind of a klutz … and a very competitive guy' at first.[3] But Hood matured – as all students do – and in the end, his work became better known than his mentor's. Indeed, a lot of what happened in the next five decades of human biology is reflected in Hood's long career. He had a knack of seeing where things were going a little ahead of others. Age and experience sometimes rob a person of their youthful drive, but this never seems to have happened to Hood.

In 1970, Hood joined Caltech's faculty himself, after spending three years at the National Institutes of Health as was required by a Vietnam War policy. He had thought about joining Harvard or Stanford, but decided that Caltech was where he could best live out Dreyer's advice.[4] He wanted to be at the frontier of biology, think about which new kinds of instruments could have a big impact there, and then set about building them. He realised that the chemical analysis of molecules which make up living things – proteins and genes – was very slow and usually done by hand. Automating these processes, he figured, would be transformative.

He began with proteins because he already knew a lot about them. Each type of protein is made up of a unique sequence of building blocks, called amino acids, linked together in a chain. There are twenty amino acids, but we still don't know how many kinds of protein they make in the human body. There are certainly more than 20,000, but if subtle variants are included, there could be billions.[5] Understanding any protein often begins with establishing the sequence of these building blocks, which provides clues to its job in the body, especially if it's similar to another type of protein whose function we already know. Identifying a protein's sequence is also a crucial step in being able to isolate the gene which is responsible for the protein's creation.

In 1981, Hood and his colleagues – most notably lab member Mike Hunkapiller – announced that they had built an instrument for sequencing proteins.[6] In a reaction chamber, individual amino acid building blocks are chopped off the protein, one at a time, allowing them to be chemically identified. Their machine proved far more reliable than any pre-existing method, and soon Hood's lab received countless important samples to analyse: hormones, nervous system receptors, blood factors, immune cell secretions, and much else.[7] Among them, in 1983, Hood's lab analysed a type of protein called prion protein.[8] Several neurodegenerative diseases in humans and animals can be caused by misshapen prion protein, including what's come to be known as 'mad cow disease', but, far from being proven, this was a radical notion at the time. Proteins were not thought capable of causing an infectious disease on their own, but only as part of something containing genetic material, as in a virus, for example.[9] Identifying the prion protein sequence on behalf of their collaborator,

Stanley Prusiner, was an important step towards this new understanding, which eventually earned him a Nobel Prize in 1997.[10] Thanks to their work, we now know that prion proteins are abundant in the human body, and that wrongly shaped prion proteins cause problems by clumping in the brain. This can be infectious, because misshapen prion protein can trigger normal prion protein to disfigure, creating new infectious particles and propagating disease in a chain reaction.[11]

Despite the obvious importance of this advancing science, when Hood pitched the protein analyser – as well as his vision for other instruments – to nineteen companies, none was interested. Perhaps this shouldn't seem so surprising. Many scientists who pioneer new tools or medicines tell stories of how they struggled at first to get others interested. (Perhaps the quintessential example of this problem was experienced by Steven Sasson, whom Prusiner met when they both received a National Medal from President Barack Obama in 2009. In 1975, Sasson built the world's first portable all-in-one digital camera, and also found a way to display pictures on a TV screen, but when he demonstrated this to his bosses at Kodak, they dismissed his invention, as they couldn't understand why anyone would want to see their photos on a TV screen.[12] Prusiner and Sasson must have hit it off, because Prusiner begins his own autobiography with what happened to Sasson.)[13] Hood thinks that one reason he failed to engage companies at the time was that he pitched his ideas to the wrong people: the middle managers rather than the company's CEOs or founders.[14] Eventually, though, a venture capitalist based in San Francisco provided seed money for Caltech to set up its own company, Applied Biosystems.[15] In 1982, it started selling a protein analyser, and soon became one of the world's leading companies for biotech instruments.[16]

Not resting on any laurels, Hood's lab produced another machine soon afterwards: a protein synthesiser. By chemically linking amino acid building blocks, instead of chopping them off, this machine could produce protein molecules by design. One of the first things Hood's lab used it for was to produce a sample of a protein normally made by HIV. Having a pure sample of this protein allowed scientists to work out its molecular structure, which in turn helped the pharmaceutical company Merck create a chemical – a protease inhibitor – to stop it working.[17] This turned out to be useful as an anti-HIV drug. Indeed, the use of protease inhibitors as medicines in 1996 – Merck's as well as one from another company, Abbott – marked the point at which AIDS began to be seen as an illness that could be controlled rather than one that was inevitably fatal.[18]

Clearly, Hood was right to follow Dreyer's advice; the eventual impact of new technologies can be far greater than anything envisaged at their inception. Still, what we've considered so far pales in comparison to what came out of Hood's lab next.

At the age of forty-one, Hood became the head of biology at Caltech, but only after it was agreed that he wouldn't have to attend a number of regular faculty meetings he thought a waste of time.[19] He wanted to stay focused on his lab, which grew in size to house over sixty-five people, about five times more than most academic groups. As one lab member put it: 'People from the outside see us as a big army, organised to scorch the earth. In fact, it's more like an amoeba, disorganised and moving in a lot of different directions.'[20] In 1982, Hood assembled a small team – each with different expertise – and gave them a mission: to build a machine for sequencing genes. Mike Hunkapiller, who had been in Hood's lab for many years already, was an engineer. His brother Tim Hunkapiller was a graduate student in Hood's lab at the time, with expertise in computing. One person was specifically recruited – Lloyd Smith, a chemist who knew about lasers – and Hood himself knew the relevant biology.[21]

Genes are made of DNA, a chemical chain with four building blocks: adenine, thymine, guanine and cytosine, known usually by their initials. In essence, a gene is a stretch of DNA – a string of As, Ts, Gs and Cs in any order – which amounts to a biological instruction, such as for a cell to produce a type of protein molecule. Gene sequencing, then, is the process whereby we determine the order of DNA's building blocks. This is vital for understanding what a gene does, how genes vary between each of us, and indeed the whole field of genetics.

The British biochemist Frederick Sanger had invented a manual process for sequencing genes a few years earlier.[22] In his method, the target gene was copied numerous times, not in its entirety but in fragments, each of differing and randomly generated lengths. A chemical process (called polymerase chain reaction or PCR) was used to add building blocks in the correct sequence for each fragment of DNA. But in the reaction test tubes, Sanger used a small amount of alternative versions of the four blocks, A, T, G and C, that were radioactive, and had a small modification to their chemical structure to halt the chemical process and stop any further building blocks being added. So when by chance a radioactive version of A, T, G or C was incorporated into the sequence in place of its normal counterpart, that fragment of DNA was complete with one radioactive building block at its end. By then counting how many building blocks preceded the radioactive tag, one could establish where in the entire target gene that end block was located. A fragment of the DNA that was, say, ten building blocks long and ended with radioactive A would tell us that A was the tenth letter in the overall sequence. Another fragment might reveal that the eleventh building block was T, and so on. With a sufficient number of fragments, one could eventually establish the entire sequence. In practice, the process used great slabs of acrylamide gel producing streams of black splodges on sheets of plastic, representing the positions of each building block in the sequence. It was an ingenious method, for which Sanger won a Nobel Prize,[23] but also slow, tedious and not entirely reliable. Hood knew that a machine to automate this process would be revolutionary.

Within about three weeks of Hood's team coming together, they realised that fluorescent coloured dyes would be a better way to tag DNA's building blocks, instead of radioactivity, and laser light could be used to make them glow.[24] It's hard to pinpoint who had this important idea; it arose in conversation, like so many great ideas. 'I don't exactly know how,' the team's chemist, Smith, later recalled, 'but some way or other we got the four-dye part.'[25] In any case, having the idea wasn't the hard part. It took a moment to have but three years to implement. The team had to work out the right chemistry to get coloured dyes onto DNA, test which types of dyes worked best, engineer an instrument which could read off the colours, develop algorithms for compiling the raw data into gene sequences, and much more besides. Eventually, in 1986, the team announced the world's first automated gene sequencer.[26] At a press conference, Hood said that this machine would have an enormous impact on our understanding of any number of diseases, from cancer to cystic fibrosis.[27] It might have sounded like hyperbole, but in fact he wasn't wrong.

After the press conference, some of the team were upset because Hood didn't mention any of them by name, which might have seemed as if he was taking the credit for himself.[28] Another problem was that the machine in Hood's lab was really just a prototype. To produce a reliable version, Applied Biosystems had to troubleshoot all sorts of technical issues, improving the chemistry of the process as well as the hardware. Hood himself described his lab's sequencer as a Ford Model A, rather than the Ferrari he wanted to build.[29] All in all, it took a lot of people, with all sorts of skills and experiences, to create a DNA-sequencing machine, and its development continued for many decades to come.

Shortly before Hood's press conference, around a dozen scientists gathered in Santa Cruz, California, on 24 May 1985, to discuss the prospect of sequencing all three billion building blocks of DNA which make up a single person's entire genome.[30] Hood was there, with Nobel laureate Walter Gilbert and future Nobel laureate John Sulston.[31] When everyone heard from Hood about DNA-sequencing machines, the mood swung from scepticism to confidence that the project was feasible.[32] The issue then became whether or not it would be worthwhile. Hood recalls the discussion as 'really quite heated'.[33] Hood himself was adamant they should do it, simply because this was 'the code for all human physiology and development'.[34] Having said that, everyone acknowledged the difficulty in how to account for variations in the genome between individuals. But first, to even know how much the human genome varied, a reference sequence was needed. (In the end, the reference human genome was derived from several volunteers, and we now know that an individual's typically differs from it at four to five million sites out of the total of three billion building blocks.[35]) There was also a sense that the human genome would surely someday be sequenced, once and for all time, so why not now?[36]

Later that year, Gilbert championed the idea at a more formal scientific symposium. The reaction was uproar at his estimate of the cost: three billion dollars. The audience were worried that the whole US budget for biology might be diverted to this singular endeavour. While other realms of science were used to big projects – the Hubble space telescope for astronomy and particle accelerators for physics – biology was still a small-team science.

Many scientists also argued that a lot of the human genome looked to be so-called 'junk DNA', in that it didn't seem to include an instruction to make a protein. So they questioned the point of sequencing it all. Thankfully, others had the foresight to realise that just because parts of the human genome were mysterious, they were unlikely to be useless. In fact, we now know that 98 per cent of the human genome doesn't code for protein in any typical way, but there are countless other treasures there instead: switches, for example, which turn other parts of the genome on and off as required in the body's different cells and tissues. In 2020, regions of this junk or 'dark' genome were found to encode for hundreds of small proteins which we barely understand anything about, but which are almost certain to have vital roles in human health and disease.[37]

Hood estimated that at first about 80 per cent of biologists were against sequencing the human genome.[38] Initially, he says, even the US National Institute of Health was against the project.[39] Despite this, Congress was interested. A panel of luminaries, including James Watson, argued the case until the Human Genome Project officially began in October 1990. Not every biologist leapt with joy. The *New York Times* reported a young scientist as saying, 'The fat cats are all getting the cream, while I'm sitting here starving.'[40]

Meanwhile, Hood was in trouble. Some of the faculty at Caltech simply didn't like his style of running a larger-than-usual lab while jet-setting around the globe to raise funds and promote the work. 'They hated my large lab,' Hood recalls, 'but I had to be large for all the different things needed.'[41] This backdrop didn't help when, in 1990, one of his team was found to have made up some of their published results. Hood's lab was all but stopped from working while an investigation ran. In the end, Hood himself was found not guilty and several scientists, including the future Nobel Prize winner James Allison, rallied to his support, especially for having acted quickly to help root out the problem.[42] Nonetheless, Hood was affected by the furore, and so other institutions thought he might be open to a move.

During a football game, a department head at the University of Washington in Seattle told Microsoft's founder Bill Gates about Hood.[43] The department head then asked Hood to give a series of three lectures in Seattle, and invited Gates to attend. The game plan worked: Hood and Gates discussed science over a three-hour dinner and, soon after, Gates funded Hood's professorship at the University of Washington.[44] Hood moved in 1992 and, after relocating, he hit the zeitgeist once again.

Hood realised, as well as others, that biology was due a sweeping change. The components of living things – genes, proteins and so on – had been, or were being, successfully identified, and so a new challenge was now emerging: to discover how all these various parts come together to create the whole, be that a cell, organ or even a person. This might seem a fairly straightforward insight, but in Hood's view it would require a radical shift in how biology is done. Pulling living things apart to see what they're made of can be done by individuals or small teams. But to understand the way in which different parts interact – the dynamics, networks and feedback loops – a new level of interdisciplinary collaboration would be needed, argued

Hood: a coming together of biologists, computer scientists and mathematicians.

In setting out his views on the matter, Hood would refer to an Indian fable in which six blind men come across an elephant, which none of them had ever encountered before. The first of them reaches out his hand and feels the elephant's side. Ah, he says, we have come across a wall. The second feels the elephant's tusk and says, no, that can't be right – for sure it is a spear. The third feels its trunk and claims they have met a snake. The fourth feels its leg and decides it is a tree. The fifth has felt the ear and says it is a fan. Finally, the sixth feels the elephant's tail and knows it is a rope. The six blind men argue and, some say, start to fight.

Hood's holistic approach to working out what a cell really is, or what a disease really is – aimed at understanding a whole system, rather than its individual parts – came to be called systems biology. The idea was, and still is, somewhat fuzzy, and even at the time it was perhaps not entirely new. As Hood himself readily acknowledged, the idea of studying biological processes on a computer was about as old as the computer itself.[45] But what Hood did was champion the idea, pushing it to the top of the community's agenda and the forefront of people's minds. In 2000 he co-founded an Institute for Systems Biology in Seattle. Of course, a lot of politics and hoo-ha was needed to get such a thing set up. Hood put millions of dollars of his own money into it, wealth he had gained from the companies he had co-founded, as well as prizes and patents.[46] Soon afterwards, systems biology became fashionable everywhere. A report by UK academic bodies in 2007, for example, recommended an investment of £325 million in systems biology.[47]

For Hood, all of this – from developing DNA-sequencing machines to the Human Genome Project and the rise of systems biology – has culminated in a new way of tackling disease. He calls it P4 medicine: Predictive, Preventive, Personalised and Participatory. 'The future of medicine is going to be very different from what we had in the past,' Hood said at a scientific meeting in September 2018.[48] He argues that dramatic increases in testing and computational capacity will make available for all of us a dense 'cloud' of personalised biological and medical data online, accessible from anywhere, which can be scrutinised to reveal our body's state of health.

For a pilot study of this idea, Hood failed to win funds from the National Institutes of Health but, characteristically, raised enough himself through philanthropic donations to go ahead anyway. One hundred and eight people had their whole genome sequenced, and had their blood, saliva, urine, and stool samples analysed every three months. (The project was somewhat like Elinav and Segal's, discussed in the previous chapter, but Hood's team examined a wider array of health issues.) Hood likened the data cloud they produced for each individual as an unprecedented deep dive into their well-being, as revelatory about a person's body as the Hubble telescope was for the universe.[49] From this, everyone received recommendations on how to improve their health. Hood participated himself, and learnt that he should increase his intake of vitamin D, because his body had a deficiency in being able to take it up. Without this, he thinks, he would be likely to get osteoporosis or Alzheimer's.[50]

Some have argued that standard medical check-ups could have detected most of the issues identified in this expensive pilot study. Hood's counter-argument is that actionable outcomes will increase in sophistication as more and more data is integrated. As we learn to read a person's data cloud more carefully, he argues, it will reveal unexpected things about what that individual should do to be healthy. While twentieth-century medicine tried to fight diseases after they arose, Hood says, twenty-first-century medicine is going to focus on catching diseases before they even happen.[51] He's right in at least one way: that it's getting easier and easier to obtain the data. It once took years and hundreds of millions of dollars to sequence a human genome. Now it takes a few hundred dollars, or less, and can be done in a single day.

In 2015, Hood co-founded a company, Arivale, to analyse a person's genes, stool, blood and saliva and report back to them on their state of health. But the cost of running so many tests

– a few thousand dollars – was more than people were willing to pay and the company collapsed four years later.[52] The general idea, however, is taking shape elsewhere. Another genome pioneer, Craig Venter, co-founded Human Longevity Inc., which also aims to use genetic sequencing and a suite of other tests for the early detection of health problems.[53] Gene-testing companies like 23andMe have signed up millions of people wanting to learn about their health or ancestry.

To be clear, others have had similar ideas to Hood along much of this journey. But among the cast of a thousand scientists, Hood has played one of the leading roles in bringing us to this point: where the human body's genetic blueprint is in hand, and countless companies are chasing ways of using it. Anyone's life takes on a hurtling pace when distilled to a few pages, but for Hood things really did move fast. As a result, a huge amount of information about our biology, and specifically our genes, has been amassed, and more is on the way. Now we must find ways to understand what it all means for who we are and how we're going to live.

Two people with the same genes – identical twins – are not the same person; everyone is more than a set of genes. Even so, *something* of us comes from our genetic inheritance. There is, at some level, such a thing as an inborn character, even if we struggle to define what that is. This is why genetic science has long been heralded as the gateway to resolving one of humankind's oldest debates – the clichéd argument of nature versus nurture, central to understanding our individuality and our identity – by at last revealing what is genetic and what isn't. But the reality uncovered thanks to the work of Hood and many others has turned out rather differently: the deeper we dig into our genetic inheritance, the messier and messier things look.

Personal character traits have turned out to be especially hard to find in our genetic code. Take intelligence, for example. As we mentioned in Chapter Four, there's a huge problem in knowing what intelligence really is and how to measure it. But even if it's defined in some narrow way, such as a score in an

IQ test, or the level of a person's educational achievement, we have yet to discover genetic variations which have a big effect on these measures (specific illnesses aside). Rather, hundreds of gene variations each appear to have some small effect.[54] This means that traits like 'intelligence' are not coded for in any simple way. Immense diversity and subtlety are the inevitable outcome of a coding system with hundreds of variables.

This has vastly complicated any attempts we might consider making at genetic interventions in humans. As we've seen in Chapter Two, with IVF, genetic screening of embryos and gene-editing tools involving CRISPR, the technology now exists to ensure that the burden of a disease-causing gene is not passed on to children. As the CRISPR pioneer Jennifer Doudna has said, 'We may be near the beginning of the end of genetic diseases.'[55] But the complexity of genetics means that even if, technical problems aside, a suite of several genetic variations could be selected for in a single embryo, there are almost certainly other traits which would be inadvertently affected as well by any intervention, resulting in knock-on effects that are almost infinitely far-reaching and impossible to fathom. This means it is implausible to select or alter an embryo to take on a particular character trait such as 'intelligence' because, however it's defined, this kind of thing isn't written in our genes in any simple way.

Instead, where the impact of genetic science – in combination with the kind of big-data analysis that Hood helped pioneer – has perhaps been felt the most is in the treatment of cancer. In 2013, the film star Angelina Jolie wrote an article for the *New York Times* about how and why she had had both of her breasts removed. Jolie had lost her mother, grandmother and aunt to cancer. After finding out from a genetic test that she had a mutation in a particular gene known as *BRCA1*, she had been given an 87 per cent chance of developing breast cancer herself. 'I can tell my children that they don't need to fear they will lose me to breast cancer,' she wrote. 'I feel empowered that I made a strong choice that in no way diminishes my femininity.'[56] The article caused a media frenzy and was a watershed moment, making millions of people aware of genetic testing and its implications. Later, in 2015, Jolie had further surgery to remove her ovaries and Fallopian tubes, to mitigate her risk of ovarian cancer. Shortly before her surgery, a blood test had revealed some inflammatory markers which, she was told, could be an early sign of cancer. Again, she wrote about her experience in the *New York Times*: 'It is not easy to make these decisions. But it is possible to take control and tackle head-on any health issue.'[57]

There is rarely a single or sudden triggering event for cancer in the way that a virus or bacteria entering the body begins an infectious illness. Rather, an out-of-control group of multiplying cells, which characterises the disease, arises over some time from a complex mix of nature and nurture. We are all aware that smoking causes lung cancer, for example.[58] Yet there are also many people who have smoked all their lives and have not suffered from lung cancer, which implies that there are many other confounding factors. Likewise, all sorts of foodstuffs have been linked to cancer, providing countless press headlines, and yet most studies remain very difficult to interpret, and don't lead to governments or health agencies being able to provide clear, declarative advice. Guidelines are important – and it is certainly good advice not to smoke – but what emerges from recent discoveries about cancer is something other than a simple list of dos and don'ts, as we will see.

Back in 1986, the first gene with a big impact on a person's risk of cancer (a 'high penetrance') was identified. Variations in the gene called *RB1* were found to be linked to a rare childhood eye tumour.[59] From 1987, analysis of this gene was offered to patients and had a clear and immediate impact: children who were at risk could be screened and, if they tested negative, spared further examination of their eye under anaesthetics.[60] Throughout the 1990s, many other genes linked to susceptibility to particular cancers were identified, including the link between breast cancer and two genes: *BRCA1* and *BRCA2* (short for 'breast cancer 1' and 'breast cancer 2', usually said as 'brak-uh' 1 or 2). A woman who has inherited a problematic mutation in one of these genes has around a 70 per cent chance of developing breast cancer some time before being eighty years old, although the precise level of risk also depends on her family history and other factors.[61] This is why the prevalence of cancer in Angelina Jolie's family significantly increased her risk.

Identification of a *BRCA1* or *BRCA2* mutation in someone with breast cancer affects their life in many ways. Instead of having surgery to remove a cancerous lump, they may choose, as Jolie did, to have a mastectomy to remove one or both breasts. They may also choose further surgery, as she also did, to mitigate an increased risk of ovarian cancer. They are likely to be offered specific types of drug treatment. In some circumstances, someone carrying a *BRCA1* or *BRCA2* mutation may choose to screen embryos after IVF, so that the problem is not passed onto their children. And of course there are implications for their family members, who may have also inherited the risk.

Angelina Jolie's particular situation is not exceptional; disease-causing mutations in *BRCA1* or *BRCA2* occur in about 1 in 400 people. Still, a mutation in *BRCA1* or *BRCA2* only accounts for a small fraction, something like 5 per cent, of all breast cancer cases.[62] Other genetic variants also associate with breast cancer, but they tend to be rarer and have less impact on a person's risk than *BRCA1* or *BRCA2*. In general, the extent to which a risk of cancer is inherited depends a lot on the type of cancer. Breast cancer is fairly typical in this regard, and altogether, inherited genes and family history account for around 10 per cent of cases. This means that the susceptibility to cancer of most patients who develop it could not have been predicted from birth. And it means that the majority of people with breast cancer are not going to pass on any increased risk to any children they have. Nonetheless, with the exception of cancers caused by, for example, viral infection, most cancers are genetic in origin – not as a result of inheritance but as a result of genetic mutations acquired in a person's cells during their lives. This happens as follows.

Whenever a cell divides, the newly formed daughter cell acquires DNA ever so slightly different from its parent cell. This is because when the parent cell's DNA is copied, the enzymes involved occasionally insert the wrong building block. Often, the mistake is noticed – because, for example, the two helical strands of DNA no longer match properly – and other enzymes fix the error. But these checks, so-called 'mechanisms of DNA repair', aren't perfect, and sometimes a change persists. An 'A' might get swapped for a 'G' somewhere in the sequence, for example. Roughly speaking, around ten letters in the string of three billion are altered whenever a cell divides. Almost always, this has no effect. But occasionally, a mutation may arise which, for example, stops a gene from working properly.[63] Some mutations might even cause a gene to produce excessive amounts of protein, or a version of protein with some abnormal activity. Over time, a series of mutations may accumulate which then cause a cell to lose its normal control over cell division, multiply excessively and thus become a cancer. Some types of cancer are more common than others – lung cancer is more common than brain cancer, for example – and one suggested explanation for this is that tissues and organs in which cells are most frequently turned over, such as the lung, are thereby more likely to accumulate cancerous mutations.[64] Mutation rates are increased by, for example, tobacco smoke, UV light and certain chemicals, which is one reason why these things increase our chance of getting cancer.

This is why genetic analysis of a person's cancerous cells can often help direct their treatment. To take one example: if the cancerous cells of a patient with lung cancer are found to include a mutation in the *EGFR* gene (epidermal growth factor receptor), meaning it produces a version of EGFR protein that is continuously active rather than responsive to demand,[65] they are more likely to respond to treatment with drugs called EGFR inhibitors.[66] In 2020, therefore, a consortium of researchers from 744 different research centres reported the genetic sequence of over 2,600 cancer samples.[67] Sophisticated cloud computing, terabytes of data and a host of algorithms had provided them with an unprecedented depth of analysis. Unlike previous studies of cancer, whole genomes were analysed, including the parts of the genome once called 'junk'. From this, several sequences associated with cancer were newly discovered. On average, it was found that each person's cancer contained four or five 'driver mutations' – changes to the genome which promote cancer directly by endowing cells with a special ability to multiply. Many other mutations that don't drive the cancer directly but accompany it were found to occur in patterns that could be used diagnostically. These were labelled 'passenger mutations'.[68] Perhaps most importantly, from so much information the trajectory of cancer development – the order in which gene mutations occur – could be inferred, and several mutations were calculated to occur long before any clinical diagnosis of cancer would be apparent: secret messages inside our cells, tell-tale signs of cancer beginning.

In the development of around half of all the cancers studied, early mutations were identified to take place in just nine genes.[69] In theory, if these mutations could be detected soon after they arise, then a risk of cancer could be diagnosed years – perhaps decades – ahead of when it might actually develop. This isn't possible at the moment, but the idea is being taken seriously. One way such mutations might be detected is by analysis of free-floating fragments of DNA which circulate in blood.

Cell-free genetic material in human blood has been known about since 1948, but only relatively recently have we been equipped with technology that is sensitive enough to analyse it. On the face of it, this offers a way of monitoring a person's genetic health at any moment. A problem is that we don't fully understand where this genetic material originates, so it's not entirely clear what's being monitored. In cancer patients, circulating DNA levels often increase, probably as a result of cancer cells being killed by the immune system, or through a natural turnover of cancer cells as some die off to be replaced by others. In these cases, cancer-derived circulating DNA, often still a small fraction of the total cell-free DNA in blood, contains mutations which can be used as a diagnostic tool. For a certain type of lung cancer known as 'non-small-cell lung cancer', for example, the presence or absence of a mutation in the *EGFR* gene can be detected in circulating blood DNA.[70] In the future, analysis of cell-free DNA in blood will become more sensitive. Whether or not this could eventually allow detection of cancer pre-disease is hard to say, but it's possible.

There are many other ideas on the table too. Minuscule packets of genes and proteins secreted from cells and circulating in blood – the menagerie of small vesicles released from cells that we met in Chapter One – might also hold messages of cancer arising.[71] This has already been shown to be the case in animals. In mice susceptible to pancreatic cancer, a vesicle marker could reveal a problem before it was detectable with an MRI scan.[72] This same marker is also present on vesicles in human patients with pancreatic cancer, and its levels in blood correlate with 'tumour burden' – the amount of cancer in the body. So although there remain many fundamental questions about what small cell-derived vesicles do in the body, they could be used as diagnostics of health, and perhaps pick up cancer or other illnesses before they fully develop.[73]

Deep cell-by-cell analysis of blood, as discussed in Chapter Three, may also play a role here. Some immune cells react against cells that have damaged DNA (as is the case with cancer cells), resulting in a slight adjustment in their characteristics, which might be detected. How early anything specific might be diagnosed isn't clear, but again, there is evidence for this being possible in animals. By monitoring levels of immune cell secretions in mice, scientists were able to detect tumour relapse very early.[74]

The microbiome is also likely to include early signifiers of cancer and other illnesses. These could be indirect – consequences of other changes to the body related to the cancer – but there is evidence that the composition of a person's gut microbes may in itself confer susceptibility to some types of cancer.[75] A person's gut microbiome can also be informative in guiding cancer therapies.[76] Patients can be identified as being likely to respond to certain immune therapies, for example, on the basis of the composition of their gut microbiome.[77] In mice, colonisation of the microbiome with an 11-strain mix of bacteria was able to improve the outcome of a specific type of immune therapy against cancer.[78]

In fact, all of these possibilities apply equally for the diagnosis or treatment of almost all types of disease. Although it's very rare that a single genetic mutation causes disease directly, gene variations make each of us more or less susceptible to almost every illness.[79] As we have seen, microbiome compositions have been linked with any number of diseases and, while exosomes and circulating blood DNA are less explored, they too are likely to vary across different states of health and disease.

Many other, even bolder ideas for monitoring our health and detecting disease have been proposed, such as analysis of our breath or the sweat from the palms of our hands, and an app which collects information about how we touch a phone can apparently reveal a depressive tendency.[80] How robustly any of these methods can be used in healthcare remains to be tested in properly controlled trials, not least because people are naturally so various in their bodies and behaviours. Perhaps, then, a useful approach would be to collect sufficient data about an individual over time to establish baseline levels of normal bodily function so that changes to these can be acted on.[81] Companies large and small have embraced this pursuit, and a boom in all kinds of predictive medicine is on the horizon.

Meanwhile, an overarching picture is emerging. Each of us is unique, a combination of attributes arising from our genes and upbringing, as well as what we eat, when we eat, how much we sleep, exercise or get stressed, our exposure to pollutants, pollen and bacteria, and myriad other influences. But for all this profound individuality, there is also a finite set of recurring patterns that indicate disease and provide a new framework for tackling it. We're used to thinking about cancer, for example, in terms of a simple list of things to do or not do: don't smoke, do use sunscreen, eat more or less of this or that type of food, and so on. But as well as this, we increasingly need to think about cancer and other diseases in a different way: in terms of the probability of their occurrence. In effect, all these new methods of analysing the human body will empower more and more people with the exact type of information that Angelina Jolie acted upon: our personal level of risk.

From such an ever-increasing deep knowledge of ourselves, we may feel emancipated and empowered. And yet probabilities are hard to weigh up. Jolie's actions in 2015 set the scene for what we all will face: an ever-increasing amount of personal biologic information presenting us with any number of difficult decisions. What does it mean to you if something has been identified that means your risk of developing cancer, or another illness, within the next twenty years is one in five? Would it be different if it was one in four? How about within five years instead of twenty? At what point would you decide to take a medicine as a precaution, or undergo a preventative operation, knowing that the medication or operation carries its own risks? With this information would you feel a victim? Would your sense of identity be affected?

The thing that triggered me to write this book was the following story. While at the checkout in a clothing store, Ruby received a call on her mobile phone. She recognised the woman's voice as the genetic counsellor she had seen once before, and asked if they could call back in five minutes. Ruby paid for her clothes, went to her car, and waited alone. Something about the counsellor's voice gave away what was coming.

They called back about ten minutes later and said Ruby's genetic test results had come in, and that she did indeed carry the mutation they had been looking for. Ruby had inherited a faulty gene from her father, the one that had caused his death aged thirty-six from a connective tissue disorder that affected his heart. It didn't seem the right situation in which to receive such news but, then again, how else could it happen? Ruby vaguely recalls that the counsellor then said she would refer her to other specialists. And that she should also consider whether or not her children should be tested. The phone call lasted five minutes or less. The counsellor asked if Ruby had any questions, but she couldn't think of anything to say. She rang off, called her husband, and cried. The main thing she was upset about was the thought of her children being at risk.

Over the next few weeks, she googled for information, read papers, and tried to become an expert patient in what was a rare genetic disorder. She couldn't find out much and, not being a scientist herself, it was hard for her to tell which, if any, of the information was reliable. She learnt that a link between mutations in this particular gene and connective tissue problems had only recently been discovered. A few years earlier and this disease didn't exist, at least not as anything named.

Over time, some details emerged. Nobody had ever seen her own family's particular mutation in anyone else. So that meant it was very hard to know what to make of her situation. Her risk of a heart problem was surely increased, but nobody could say by how much. She knew about Angelina Jolie. But this felt different. Jolie was told she had a very high chance of breast cancer, and Ruby's situation was much less clear.

From the car park phone call, it took over six months for Ruby to be seen by any other medical professional. She saw a cardiologist first, followed quickly by a series of other specialists, as each appointment seemed to trigger a chain of others. The outcome was that Ruby would have regular body scans, and she began to take medication to lower her blood pressure, which she was told to do as a precaution for the rest of her life. She was also told to avoid anything that would cause her body to suddenly jolt. The vagueness of what this meant in practice became another source of worry. Should she should carry on playing basketball, for example? The decision was left to her, and every opinion was available online. Other small things also had a big effect. She had always loved holidays abroad, but now travel insurance became exceptionally hard for her to get, because nobody knew how to categorise her body.

Ruby told me this story herself. Ultimately, the difficulty she faced is that, at the edge of science, there's so much uncertainty. Ruby believes that it is definitely better to have been informed of her genetic inheritance, because in her case there are things she can do to lower the risk of there being a real problem. But it took a long time for her to come to understand that she was not actually ill. She was only at risk of being ill. In fact, nothing had actually changed; she had only become aware of a possible future. Still, once you know something like this, it's impossible to live as if you didn't.

A holy grail of science and medicine is to stop disease. Especially to stop it before it even begins. For some illnesses, this has been achieved already – with vaccines, clean water and improved sanitation. Now, with the patterns and codes behind how the human body works opening up to us, new ways of doing this are emerging. We are compelled to seize this new opportunity, yet in practice there are challenges and unintended consequences to contend with.

I've heard it said that you can't write a book unless you have to write a book. The reason I had to write this one – and especially Ruby's story – is that every one of us is susceptible to some disease or other to some extent. So as science progresses and we learn more and more about ourselves, we will surely all find ourselves in a similar situation to Ruby one day; awash with estimates and probabilities which play games with our mind and our identity, and require us to make difficult decisions about our health and how we live. I wrote this book not because I have the answer to how we deal with this. It's simply that understanding the background may help.

What all of this – the secret body – means for our lives is still very much in the balance. We are drowning in data which says that every one of us is sub-optimal. Or that every one of us is special. It depends on how you look at it. We are not merely our genes, our cells, our microbiome or our brain. We are all these things. We are more than all these things. What emerges is this: that our body matters, but it isn't everything that matters. How we see ourselves and others – the story we live in, and the philosophies we live by – are just as vital. Context is everything, a little of which is all any book can ever hope to give.

7 What it all Means

'Now', said the Voice, ... 'Am I imagination?'

H. G. Wells, *The Invisible Man* (1897)

Babies are now routinely born by IVF, organ transplants have become common, and cancer survival in the UK has roughly doubled in recent years – but all these achievements are nothing to what's coming. Progress in human biology is accelerating at an unprecedented rate. On the horizon now are entirely new ways of defining, screening and manipulating health, completely new ideas about cells, bacteria, diet and the human brain, and any number of ideas for how babies can be born. We are not simply meandering from the last few decades into the next few decades, with a few details being tweaked here and there; we are at the cusp of a revolutionary time in virtually every aspect of human biology.

Huge global upheaval from the biological manipulation of nature has occurred before. When humankind began to domesticate crops, livestock and pets, this eventually led to the development of cities, complex economics and political hier-archies – none of which could have been predicted, nor were the aim at the outset. In turn, these led to other issues such as the spread of infectious diseases and problems to do with money and power. Likewise, it's impossible to predict how today's new advances in human biology will impact our lives in a hundred or a thousand years from now. The journey has only just begun, and for where we're going next, there's no map. But by bringing together the findings from these six frontiers of human biology it has hopefully become clear that this scientific revolution is going to affect us in a very different way to previous ones.

Whereas the agricultural, industrial and digital revolutions have affected our environments and societies, new human biology equips each of us individually, physically and psychologically with new powers, and each of us will need to decide for ourselves if and when to deploy them. In the near future we will have to decide, for example, whether or not to take nutritional advice from an algorithm that has analysed the components of our own stool and blood. Our body's cells will have been scrutinised, especially in our immune system, and the results will show that we are prone to develop this or that particular problem. We will then need to decide whether or not to take a variety of precautionary measures. Deeper than that, from the knowledge of being at risk of something named, our sense of self will change.

Soon, we will also face the opportunity to boost our cognition, with the knowledge that others almost certainly will. Surely this too will shift our sense of self, not least because success in work, or educational achievement, will change its meaning if drugs can be used to affect those things. We will also need to decide whether or not to use drugs which can relieve depression or hack our emotions in some way. And as if these decisions aren't difficult enough for ourselves, we will need to make choices on behalf of our children too, including before they are born. All of these dilemmas will affect us enormously – across the course of our lives – and each of us will have to engage directly with science to navigate them.

And yet, even now, scientific information struggles to be heard or understood within a clatter of hashtags and retweets. Uncertain of who to trust and perhaps sceptical of science's presumed authority, parents may, for example, reject advice to have their children vaccinated. Graphs and data can help explain things, but what's also needed is a deeper understanding of how science works. Never before has widespread discussion of scientific ideas – especially new science of the human body – been more important, for society and each of us individually.

When all of this is looked at together, something else becomes apparent too. In Kazuo Ishiguro's novel *The Buried Giant*, a mystical fog – the breath of a she-dragon – descends over Britain's villages and alters people's memories, shifting their understanding of who they are and how they relate to one another. If there's something even remotely akin to this in real life, it is perhaps the spread of scientific knowledge. In Ishiguro's novel, people who were once enemies are reconciled when the mist muddles their remembrances of past events. By revealing our basic nature, making plain who and what we are, science is also a force which can bring us together. Yet, where Ishiguro's dragon mist brought loss and forgetfulness, scientific knowledge helps in the opposite way: by opening up new ways of seeing ourselves.

Science is often perceived as bringing us exactness and precision, which of course it does do in many ways, or else we would not have cars or iPhones. But the deeper we study the human body, the more we find ourselves not to be so precisely defined. We are fundamentally dynamic, plastic and entwined with a universe of cells which are not even human. Almost magical. This undoes any number of historically divisive and dogmatic views. Racial purity is nonsense, for example. There is a rough geographical distribution of gene variations across the globe, but the borders are fuzzy and everyone is related to everyone else, not as a metaphor but as a fact. Science is vital to understanding – and appreciating – this human diversity. Increasingly, it is revealing that diversity to be even greater and more profound than we had realised.

And yet, as science propels us into this new epoch, we must take care to guard against this knowledge and its application becoming a new source of division in the world at large. While writing this book with all its highfalutin ideas and opportunities, it kept coming back to me that much of the world still lacks basic sanitation. While it is surely a noble quest to understand how billions of cells work together to create what we are, and many of the things we have discovered will lead to new cures and treatments of human disease, we must not be so dazzled that we miss seeing the problems in global access to healthcare. Science is a relentless pursuit of more, but it must not be for only the few.

At the outset, I mentioned that there are many ways of thinking about Leonardo da Vinci's painting the *Mona Lisa*. Yet the real thing, hanging in the Louvre museum in Paris, is so surprisingly small. All the fuss contained in a small rectangle, 77 x 53 cm. Each of us is also very small – less than a dot in an unfathomably vast universe – and yet each of us has a magnificence which is impossible to take in. Travelling inside ourselves is the greatest, most soulful, and probably the most important, adventure we have ever begun.

Acknowledgements

I am especially grateful to everyone I had the privilege of interviewing for this book, including: Eric Betzig, Moshe Biton, Paul Brehm, Ali Brivanlou, Martin Chalfie, Matthew Cobb, Sheena Cruickshank, Eran Elinav, Paula Garfield, Jerome de Groot, Muzlifah Haniffa, Leonore Herzenberg, Leroy Hood, Susan Kimber, Jack Kreindler, Jeff Lichtman, Jennifer Lippincott-Schwartz, Elizabeth Mann, Ashley Moffett, Werner Müller, Paul Norman, Luke O'Neill, Jordan Orange, Berenika Plusa, Jon Price, Aviv Regev, Andrew Sharkey, Sachi Shimomura, Elizabeth Simpson, Christoph Wülfing, Magdalena Zernicka-Goetz.

I am also indebted to many others who helped me address specific issues, including: Murray Buchanan, George Church, Michael Dustin, John Hammer, Gareth Howell, Konrad Krzewski, Kathleen Nolan, Seth Scanlon, Santiago Zelenay. I am also hugely grateful to those who read early versions of some or all the text, including: Judith Allen, Moshe Biton, David Brough, Matthew Cobb, Andrew Doig, Eran Elinav, Khodor Hazime, Matthew Hepworth, Gareth Howell, Viki Male, James Nicholls, Camille Rey, Karoliina Tuomela, Jonathan Worboys. I'm also grateful to anonymous peer reviewers who also provided helpful comments. Needless to say, any remaining errors in this book are my responsibility alone. I also thank all the members of my research team who have guided my thinking over many years.

My editor at the Bodley Head, Will Hammond, has been exceptionally supportive and has had a huge influence on the overall shape of this book as well as its details. I am also deeply grateful to Caroline Hardman, my literary agent, who has been enormously helpful in all sorts of ways for all three books I have written. At Princeton University Press, I thank Alison Kalett for her vital feedback and comments, as well as Christie Henry for championing this book at the outset. Graham Coster copy-edited the text, and managed to solve problems with the endnotes that threatened to stretch to the end of days. And thanks to Lauren Howard and Mia Quibell-Smith at the Bodley Head for also being a huge help in the final stages of preparing and launching this book.

On a different level, I thank my parents, Marilyn and Gerald Davis, and all my family, for their enduring support. Above all, I thank my wife Katie, and our children Briony and Jack, for sharing the journey.

Notes

1. Super-resolution Cells

1 It's not precisely clear when or how a bona fide microscope was invented. One reason why it's hard to be sure who did this first – and at least four different Dutch instrument makers have been credited – is that once out there the idea can be quickly copied. Also, tools to visually magnify objects long predate actual microscopes. Indeed, ever since humankind first used water-filled spheres to see small objects, in ancient Greece and elsewhere, we have enjoyed using tools with the power of magnification. For more details on this see, for example: Bardell, D., 'The Biologists' Forum: The invention of the microscope', *BIOS* 75 (2004), pp. 78–84.

2 Samuel Pepys's diary, which he kept from 1660 to 1669, is widely regarded as an important insight into everyday life in London at the time. His diary entry for 21 January 1665 includes the line: 'Before I went to bed I sat up till two o'clock in my chamber reading of Mr Hooke's *Microscopical Observations*, the most ingenious book that ever I read in my life.'

3 Hooke deliberately linked his microscopic observations with everyday life. For example, he showed a louse clinging to a human hair, rather than showing a louse on its own.

4 Leeuwenhoek sent a letter to the Royal Society in 1677 describing his discovery. At the time, he didn't think that sperm were linked to heredity, rather that they were some kind of animal or parasite. Later, he did consider sperm as being vital to creating offspring, but he mistakenly thought they contained the complete embryo. It wasn't until much later, after cell theory was developed in the 1840s, that a modern view of sperm and egg cells began

to emerge. This is discussed in detail in Cobb, M., *The Egg and Sperm Race* (Free Press, 2006).

5 Lauterbach, M. A., 'Finding, defining and breaking the diffraction barrier in microscopy – a historical perspective', *Optical Nanoscopy* 1, (2012), 8.

6 Abbe's famous paper published in 1873 is celebrated for explicitly stating that the resolution of a microscope is limited by the wavelength of light. However, many other scientists also contributed to our understanding that microscopes have a fundamental limit. In 1874, the German physician and scientist Hermann von Helmholtz published a detailed mathematical analysis which came to the same conclusion as Abbe, and von Helmholtz stated that he finished his analysis before he was aware of Abbe's work.

7 The principle behind an electron microscope is similar to a regular light microscope, except that beams of electrons replace light. Higher resolution is achieved in an electron microscope because the wavelength of electrons is much shorter than light. Coiled electromagnets or solenoids replace glass lenses to direct beams of electrons, which are then detected with photomultiplier tubes.

8 Chang, K., 'Osamu Shimomura, 90, Dies; Won Nobel for Finding a Glowing Protein', *New York Times*, 24 October 2018.

9 Shimomura, O., 'The discovery of aequorin and green fluorescent protein', *Journal of Microscopy and Ultrastructure* 217 (2005), pp. 1–15.

10 Interview with Sachi Shimomura, 18 February 2019.

11 Davenport, D. and Nicol, J. A. C., 'Luminescence in Hydromedusae', *Proceedings of the Royal Society of London*, Series B – Biological Sciences 144 (1955), pp. 399–411.

12 Olson, E. R., Martin, J. G., Anich, P. S. and Kohler, A. M., 'Ultraviolet fluorescence discovered in New World flying squirrels (Glaucomys)', *Journal of Mammology*, gyy177 (2019), https://doi.org/110.1093/jmammal/gyy1177.

13 Interview with Paul Brehm, 28 January 2019.

14 Email correspondence with Sachi Shimomura, 21 February 2019.

15 Shimomura, O., Shimomura, S. and Brinegar, J. H., *Luminous Pursuit: Jellyfish, GFP, and the Unforeseen Path to the Nobel Prize* (World Scientific, Hackensack, New Jersey, 2017).

16 Ibid.

17 Interview with Sachi Shimomura, op. cit.

18 Shimomura, O., Johnson, F. H. and Saiga, Y., 'Extraction, purification and properties of aequorin, a bioluminescent protein from the luminous hydromedusan, Aequorea', *Journal of Cellular and Comparative Physiology* 59 (1962), pp. 223–39. This was the first report on these two proteins. The emphasis of the paper was in the isolation of the calcium-sensitive protein, named aequorin, but the presence of a green protein was also noted.

19 Many other scientists were also involved in the isolation of GFP and its development as a tool. James Morin and John 'Woody' Hastings at Harvard University, for example, studied many luminescent organisms and proteins, and they coined the name green fluorescent protein in this paper: Morin, J. G. and Hastings, J. W., 'Energy transfer in a bioluminescent system', *Journal of Cellular Physiology* 77 (1971), pp. 313–18.

20 The talk was by the neurobiologist Paul Brehm, then at Tufts University. Brehm did his PhD research studying the luminescence of marine life under James Morin, one of the pioneers in isolating and understanding GFP. While at the State University of New York, Brehm had also collaborated with Shimomura in studying the luminous brittle star *Ophiopsila* (a type of animal closely related to starfish).

21 This was the *C. elegans* worm. Chalfie had worked as a postdoc with Sydney Brenner as the first neurobiologist to study this worm. He wanted to understand how these small worms were able to respond to being touched, and thought that maybe GFP could report where touch-sensitive genes were being switched on in the animal.

22 Martin Chalfie, Nobel Lecture, 2008. Martin Chalfie delivered his Nobel Lecture on 8 December 2008, at Aula Magna, Stockholm University. Available online at: https://www.nobelprize.org/prizes/chemistry/2008/chalfie/lecture/

23 Prasher, D. C., Eckenrode, V. K., Ward, W. W., Prendergast, F. G. and Cormier, M. J., 'Primary structure of the Aequorea victoria green-fluorescent protein', *Gene* 111 (1992), pp. 229–33.

24 In Chalfie's lab, it was the graduate student Ghia Euskirchen and lab technician Yuan Tu who first expressed the GFP gene in bacteria and worms respectively, so that they fluoresced green.

25 Chalfie, M., Tu, Y. and Prasher, D. C., 'Glow Worms – A New Method of Looking at *C. elegans* Gene Expression', *Worm Breeder's Gazette* 13 (1993), p. 19.

26 Chalfie, M., Tu, Y., Euskirchen, G., Ward, W. W. and Prasher, D. C., 'Green fluorescent protein as a marker for gene expression', *Science* 263 (1994), pp. 802–5.

27 Interview with Martin Chalfie, 21 January 2019.

28 Frederick Tsuji at Scripps Institution of Oceanography also published that GFP could be expressed in bacteria shortly after Chalfie and Prasher's celebrated paper in 1994.

29 Bhattacharjee, Y., 'How bad luck and bad networking cost douglas prasher a nobel prize', *Discover*, April 2011.

30 Chang, K., 'Man who set stage for a nobel now lives a life outside science', *New York Times*, 16 October 2008.

31 Email correspondence with Martin Chalfie, 26 January 2019.

32 Zimmer, M., *Illuminating Disease: An Introduction to Green Fluorescent Proteins* (Oxford University Press, 2015).

33 Sherwell, P., 'The scientist, the jellyfish protein and the Nobel Prize that got away', *Telegraph*, 11 October 2008.

34 Chang, 'Osamu Shimomura, 90, Dies ...', *op. cit.*

35 Interview with Eric Betzig, 6 February 2019.

36 In fact, like so many scientific stories, the invention of the laser is exceptionally complicated. See for example, Nick Taylor's *Laser: The Inventor, the Nobel Laureate, and the Thirty-year Patent War* (Simon and Schuster, New York, 2000). The Nobel Prize in Physics 1964 was awarded for the invention of the laser, and surrounding work, to Charles Townes, Nicolay Basov and Aleksandr Prokhorov. But others, including Theodore Maiman and Gordon Gould, are also often credited with the invention of the laser. Battles over patents lasted for around thirty years, making the discovery of the laser one of the most legally disputed scientific inventions ever.

37 Betzig, E., 'Proposed method for molecular optical imaging', *Optics Letters* 20 (1995), pp. 237–9.

38 Eric Betzig, Nobel Lecture, 2014. Eric Betzig delivered his Nobel Lecture on 8 December 2014, at Aula Magna, Stockholm University. Available online at: https://www.nobelprize.org/prizes/chemistry/2014/betzig/lecture/

39 Ibid.

40 Ibid.

41 Email correspondence with Eric Betzig, 6 February 2019.

42 Interview with Eric Betzig, 6 February 2019.

43 A photo of their homebuilt microscope in Hess's living room is shown in Eric Betzig's Nobel lecture.

44 Patterson, G. H. and Lippincott-Schwartz, J. A., 'Photoactivatable GFP for selective photolabeling of proteins and cells', *Science* 297 (2002), pp. 1873–7.

45 As Betzig readily acknowledges, there were others before him who had developed instruments which beat Abbe's law. For example, Eric Ash (who incidentally did his PhD work with the inventor of holography, Denis Gabor) developed scanning near-field microscopy, which achieves a resolution well below the limit imposed by Abbe's law. The original theory behind this type of microscopy was published in 1928 by the Irish scientist Edward Synge. Decades later, in 1972, Ash demonstrated the concept using 3cm microwaves (Ash, E. A. and Nicholls, G., 'Super-resolution aperture scanning microscope', *Nature* 237 (1972), pp. 510–2.) Betzig and others built upon Ash's work and developed the method, which is still widely used today. Crucially, however, this type of microscopy can only study samples at an exceptionally shallow depth, probing only the uppermost surface of cells, for example. It was frustration with this technique, as well as Betzig's sense that, where he worked in Bell Labs, basic science was going to be less valued than it had been, which led to him quitting science for a while.

46 George Patterson, working in Jennifer Lippincott-Schwartz's lab, created photoactivatable GFP. He did this because he wanted to study the movements of proteins between different compartments inside cells. He realised that if you could switch on the green glow of GFP in one compartment of the cell, you could then follow where else that protein moved to inside the cell. Until they met Betzig, they didn't realise that this version of GFP could also enable super-resolution microscopy.

47 Interview with Jennifer Lippincott-Schwartz, 5 March 2019.

48 Ibid.

49 Betzig, E. *et al.*, 'Imaging intracellular fluorescent proteins at nanometer resolution', *Science* 313 (2006), pp. 1642–5.

50 Betzig, Nobel Lecture, 2014, *op. cit.*

51 Rust, M. J., Bates, M. and Zhuang, X., 'Sub-diffraction-limit imaging by stochastic optical reconstruction microscopy (STORM)', *Nature Methods* 3 (2006), pp. 793–5.

52 Zhuang's formal paper was published first, on 9 August 2006, with Betzig and Hess's one day later. Betzig and Hess had first

presented their findings at a meeting in the National Institutes of Health in April 2006, but the peer-review process took some time, with one reviewer requesting that they specifically correlate their new microscope images with the same samples also imaged by electron microscopy, something that is technically very challenging. A full profile of Xiaowei Zhuang is published here: Vilcek, J. and Nair, P., *Proceedings of the National Academy of Science of the USA*, 117 (2020), pp. 9660–9664.

53 Hess, S. T., Girirajan, T. P. and Mason, M. D., 'Ultra-high resolution imaging by fluorescence photoactivation localization microscopy', *Biophysical Journal* 91 (2006), pp. 4258–72.

54 Stefan W. Hell – Biographical. NobelPrize.org. Nobel Media AB 2014. Available at https://www.nobelprize.org/prizes/chemistry/2014/hell/biographical/

55 A stepping stone to Stefen Hell's most famed breakthrough was his development of what came to be called the 4Pi microscope, which he successfully demonstrated while working with Ernst Stelzer in 1994. This microscope uses two objective lenses, one either side of the sample, which improves the axial resolution of the microscope.

56 There are many ways to create a tube-shaped laser beam. In practice, this often involves passing the laser light through a special glass plate. The details of how this work are complex, and in fact there many research papers testing which method works best for super-resolution microscopy.

57 There's a subtlety here which you may have thought about. According to Abbe's law, a sharp ring of light will also be blurred. This is true, but in practice the intensity (and other factors) of the doughnut-shaped laser can be adjusted so that nearly all molecules are switched off in the outer edges of where the first laser hits. There are also several variations that can improve this technology. One common approach is called gated STED. In this set-up, the light is collected after a short delay, which helps make sure the outer-edge molecules have had time to be darkened.

58 Hell, S. W. and Wichmann, J., 'Breaking the diffraction resolution limit by stimulated emission: stimulated-emission-depletion fluorescence microscopy', *Optical Letters* 19 (1994), pp. 780–2.

59 Klar, T. A., Jakobs, S., Dyba, M., Egner, A. and Hell, S. W., 'Fluorescence microscopy with diffraction resolution barrier broken

by stimulated emission', *Proceedings of the National Academy of Sciences of the USA* 97 (2000), pp. 8206–10.

60 Lippincott-Schwartz, J., 'Profile of Eric Betzig, Stefan Hell, and W. E. Moerner, 2014 Nobel Laureates in Chemistry', *Proceedings of the National Academy of Sciences of the USA* 112 (2015), pp. 2630–2.

61 Dickson, R. M., Cubitt, A. B., Tsien, R. Y. and Moerner, W. E., 'On/off blinking and switching behaviour of single molecules of green fluorescent protein', *Nature* 388 (1997), pp. 355–8.

62 Another super-resolution technology – not directly recognised by the Nobel Prize but widely used nonetheless — involves illuminating the sample with fine stripes of light, whose position and orientation are changed a number of times, such that computational calculations can extract high-resolution information from the emitted light. This method, called structured illumination microscopy (or SIM), beats Abbe's law, but only to a modest extent compared to the technologies developed by Betzig, Hell and their colleagues. An advantage of SIM, however, is that it is fast and well suited for long-term imaging of live cells. Mats Gustafsson invented SIM microscopy, but sadly he died from brain cancer in 2011 at the age of 51. In a tribute published in *Nature Methods* 8 (2011), p. 439, Betzig said: 'He didn't write many papers, but every paper was a bible for that method.'

63 Betzig, Nobel Lecture, 2014, op. cit.

64 Interview with Eric Betzig, 6 February 2019.

65 Dreifus, C., 'Life Over the Microscope', *New York Times*, 1 September 2015.

66 The words here reflect what he said in his talk on the day, which differs very slightly from the official transcript of his lecture. The lecture is available here: https://www.nobelprize.org/prizes/chemistry/2014/betzig/lecture/

67 There are also other problems in the immune system of Chediak-Higashi patients, including the fact that their macrophages are not efficient at destroying bacteria.

68 Gil-Krzewska, A. *et al.*, 'An actin cytoskeletal barrier inhibits lytic granule release from natural killer cells in patients with Chediak-Higashi syndrome', *Journal of Allergy and Clinical Immunology* 142 (2018), pp. 914–27.

69 Brynner, R. and Stephens, T., *Dark Remedy: The Impact of Thalidomide and its Rival as a Vital Medicine* (Basic Books, 2001).

70 Lagrue, K., Carisey, A., Morgan, D. J., Chopra, R. and Davis, D. M., 'Lenalidomide augments actin remodeling and lowers NK-cell activation thresholds', *Blood* 126 (2015), pp. 50–60.

71 While I highlight this particular story, I am, of course, hugely indebted to all sixty-three doctoral students and postdocs who have worked in my lab to date. During every one of our weekly lab meetings, ideas flow between us collectively, each of us being influenced by others in any number of ways.

72 Interview with Lippincott-Schwartz, *op. cit.*

73 Hirschberg, K. *et al.*, 'Kinetic analysis of secretory protein traffic and characterisation of golgi to plasma membrane transport intermediates in living cells', *Journal of Cellular Biology* 143 (1998), pp. 1485–1503.

74 'The Microscopists interviews Jennifer Lippincott-Schwartz' (Howard Hughes Medical Institute), an interview conducted by Peter O'Toole, 27 August 2020, available online here: https://youtu.be/XiofXaNnMZQ

75 Nixon-Abell, J. *et al.*, 'Increased spatiotemporal resolution reveals highly dynamic dense tubular matrices in the peripheral ER', *Science* 354 (2016), aaf3928.

76 Xu, K., Zhong, G. and Zhuang, X., 'Actin, spectrin, and associated proteins form a periodic cytoskeletal structure in axons', *Science* 339 (2013), pp. 452–6.

77 In Zhuang's Breakthrough Prize Symposium talk, on 5 November 2018, at UC Berkeley, she was asked directly why nobody saw these structures earlier using electron microscopy. She replied that she thought it was because detergents used to help stain the protein which makes up the rings would disrupt its structure. Her talk is available online here: https://www.youtube.com/watch?v=KmIaUQa-QyQ

78 Sigal, Y. M., Zhou, R. and Zhuang, X., 'Visualising and discovering cellular structures with super-resolution microscopy', *Science* 361 (2018), pp. 880–7.

79 The discovery also turned out to be lucrative: in 2019, Zhuang won a Life Sciences Breakthrough Prize, worth $3 million, sponsored by the founders of Google and Facebook and others, 'for discovering hidden structures in cells by developing super-resolution imaging'.

80 Harding, C. V., Heuser, J. E. and Stahl, P. D., 'Exosomes: looking back three decades and into the future', *Journal of Cell Biology* 200 (2013), pp. 367–71.

81 Raposo, G. *et al.*, 'B lymphocytes secrete antigen-presenting vesicles', *Journal of Experimental Medicine* 183 (1996), pp. 1161–72.

82 Valadi, H. *et al.*, 'Exosome-mediated transfer of mRNAs and microRNAs is a novel mechanism of genetic exchange between cells', *Nature Cell Biology* 9 (2007), pp. 654–9.

83 Davis, D. M., 'Intercellular transfer of cell-surface proteins is common and can affect many stages of an immune response', *Nature Reviews Immunology* 7 (2007), pp. 238–43.

84 van Herwijnen, M. J. *et al.*, 'Comprehensive Proteomic Analysis of Human Milk-derived Extracellular Vesicles Unveils a Novel Functional Proteome Distinct from Other Milk Components', *Molecular and Cellular Proteomics* 15 (2016), pp. 3412–23.

85 Boulanger, C. M., Loyer, X., Rautou, P. E. and Amabile, N., 'Extracellular vesicles in coronary artery disease', *Nature Reviews Cardiology* 14 (2017), pp. 259–72.

86 Hoshino, A. *et al.*, 'Tumour exosome integrins determine organotropic metastasis', *Nature* 527 (2015), pp. 329–35.

2. The Start of Us

1 Knight, K., 'Does my son prove babies with gene defects can cure themselves in the womb?' *Daily Mail*, 7 April 2016.

2 Interview with Magdalena Zernicka-Goetz, 11 April 2019.

3 Vogel, G., 'Pushing the limit', *Science* 354 (2016), pp. 404–7.

4 Talking to me in April 2019, Zernicka-Goetz recalls that Tarkowski didn't take on PhD students as such. Rather, he simply took people on to conduct research in his lab. Zernicka-Goetz was tasked with trying to make an embryo with components from two different rodents – between mice and bank voles, or mice and rats. This never worked; she found that a nucleus from an embryo cell couldn't survive in the cytoplasm of an embryo cell from a different species. When Zernicka-Goetz broke her arm, preventing her carrying out experiments for a while, she asked Tarkowski if she could write up her work for a PhD and he said yes.

5 Evans, M. J. and Kaufman, M. H., 'Establishment in culture of pluripotential cells from mouse embryos', *Nature* 292 (1981), pp. 154–6.

6 In December 1981, Gail Martin, at the University of California, San Francisco, also published a way to isolate and grow embryonic stem cells.

7 Evans went on to establish that embryonic stem cells could be genetically modified. And he showed that such modified cells could be re-introduced to a mouse embryo, leading to an animal being born with new genetic material. This was the key work cited for Evans winning a Nobel Prize in 2007.

8 Martin Evans, Nobel Lecture, 2007. Sir Martin J. Evans delivered his Nobel Lecture on 7 December 2007 at Karolinska Institutet in Stockholm. Available online at: https://www.nobelprize.org/prizes/medicine/2007/evans/lecture/

9 These embryos became healthy babies when implanted into the uterus of a mouse, showing that the green-glowing jellyfish protein didn't interfere with normal development.

10 Zernicka-Goetz, M. *et al.*, 'Following cell fate in the living mouse embryo', *Development* 124 (1997), pp. 1133–7.

11 Piotrowska-Nitsche, K. and Zernicka-Goetz, M., 'Spatial arrangement of individual 4-cell stage blastomeres and the order in which they are generated correlate with blastocyst pattern in the mouse embryo', *Mechanisms of Development* 122 (2005), pp. 487–500.

12 Tarkowski, A. K., 'Experiments on the development of isolated blastomers of mouse eggs', *Nature* 184 (1959), pp. 1286–7.

13 Vogel, G., 'Embryology. Embryologists polarized over early cell fate determination', *Science* 308 (2005), pp. 782–3.

14 Zernicka-Goetz, M., 'Cleavage pattern and emerging asymmetry of the mouse embryo', *Nature Reviews Molecular Cell Biology* 6 (2005), pp. 919–28.

15 Magdalena Zernicka-Goetz talking at the 2016 Childx Symposium, on paediatric and maternal health, held at Stanford University. Her talk is online here: https://www.youtube.com/watch?v=7cZhuXTvfis

16 Ibid.

17 Zhang, H. T. and Hiiragi, T., 'Symmetry Breaking in the Mammalian Embryo', *Annual Review of Cell and Developmental Biology* 34 (2018), pp. 405–26.

18 This was proven by Zernicka-Goetz and other labs, by showing that a specific profile of gene activity had been switched on in each cell. Two papers published together reported these results: White, M. D. *et al.*, 'Long-Lived Binding of Sox2 to DNA Predicts Cell Fate in the Four-Cell Mouse Embryo', *Cell* 165 (2016), pp. 75–87, and Goolam, M., et al., 'Heterogeneity in Oct4 and Sox2 Targets Biases Cell Fate in 4-Cell Mouse Embryos', ibid, pp. 61–74.

19 Magdalena Zernicka-Goetz recalled this in a talk (which I chaired) at the Hay Festival on Monday, 29 May 2017, entitled 'The Start of Life – How far should science go?'

20 Her lab team created embryos containing abnormal cells as follows. They first used a drug to produce wholly defective embryos. Then, abnormal cells isolated from these embryos were combined with other cells from a healthy embryo to create mosaic embryos.

21 Bolton, H. *et al.*, 'Mouse model of chromosome mosaicism reveals lineage-specific depletion of aneuploid cells and normal developmental potential', *Nature Communications* 7 (2016), 11165.

22 Loke, Y. W., *Life's Vital Link: The Astonishing Role of the Placenta* (Oxford University Press, 2013).

23 Vestre, K., *The Making of You: A Journey from Cell to Human* (Profile Books, Wellcome Collection, 2019).

24 Yutkey, K. E. and Kirby, M. L., 'Wherefore heart thou? Embryonic origins of cardiogenic mesoderm', *Developmental Dynamics* 223 (2002), pp. 307–20.

25 Yamaguchi, Y. and Yamada, S., 'The Kyoto Collection of Human Embryos and Fetuses: History and Recent Advancements in Modern Methods', *Cells Tissues Organs* 205 (2018), pp. 314–19.

26 Leeton, J., 'The early history of IVF in Australia and its contribution to the world (1970–1990)', *Australian and New Zealand Journal of Obstetrics and Gynaecology* 44 (2004), pp. 495–501.

27 Edwards, R. and Steptoe, P., *A Matter of Life: The Story of a Medical Breakthrough* (Hutchinson, 1980).

28 Ibid.

29 A blog piece for the Science Museum, London by Connie Orbach, published on 9 July 2018, 'Jean Purdy, The Forgotten IVF Pioneer', is available online here: https://blog.sciencemuseum.org.uk/jean-purdy-the-forgotten-ivf-pioneer/

30 Brown, L., *My Life as the World's First Test-tube Baby* (Bristol Books CIC, Bristol, 2015).

31 This enormous achievement built upon countless years of endeavour. The first step, for example, reported in 1969 – fertilisation of a human egg with a human sperm in a lab dish – was already a feat less easy than it might sound, requiring, for example, a source of human eggs from ovary biopsies and then finding conditions that activate or 'capacitate' sperm, which turn out to

require a mildly alkaline environment. This was reported in Edwards, R. G., Bavister, B. D. and Steptoe, P. C., 'Early stages of fertilisation *in vitro* of human oocytes matured *in vitro*', *Nature* 221 (1969), pp. 632–5. There were several earlier reports of fertilisation being achieved between a human egg and sperm in a lab dish, for example by John Rock at Harvard University and by Landrum Shettles at Columbia University, but such claims were discredited or at least not clearly proven.

32 Rorvik, D., 'The embryo sweepstakes: The winner will be a brave new baby conceived in a test-tube and then planted in a womb', *New York Times*, 15 September 1974.

33 Edwards and Steptoe, *A Matter of Life, op. cit.*

34 Johnson, M. H., Franklin, S. B., Cottingham, M. and Hopwood, N., 'Why the Medical Research Council refused Robert Edwards and Patrick Steptoe support for research on human conception in 1971', *Human Reproduction* 25 (2010), pp. 2157–74.

35 Faddy, M. J., Gosden, M. D. and Gosden, R. G., 'A demographic projection of the contribution of assisted reproductive technologies to world population growth', *Reproductive Biomedicine Online* 36 (2018), pp. 455–8.

36 Shahbazi, M. N. *et al.*, 'Self-organisation of the human embryo in the absence of maternal tissues', *Nature Cell Biology* 18 (2016), pp. 700–8.

37 The report is formally known as the Report of the Committee of Inquiry into Human Fertilisation and Embryology. It was published in 1984 and is not to be confused with another report, also colloquially known as the Warnock Report, published in 1978, officially the Report of the Committee of Enquiry into the Education of Handicapped Children and Young People. Warnock's 1978 report was also pioneering and hugely influential, leading to legislation for educational inclusion and changing the way society treated disability.

38 Hyun, I., Wilkerson, A. and Johnston, J., 'Embryology policy: Revisit the 14-day rule', *Nature* 533 (2016), pp. 169–71.

39 Ditum, S., 'Public intellectuals have never been more vital. Let Mary Warnock be a guide', *Guardian*, 24 March 2019.

40 Warnock, M., *A Memoir: People and Places* (Duckworth, 2000).

41 Hurlbut, J. B, *et al.*, 'Revisiting the Warnock rule', *Nature Biotechnology* 35 (2017), pp. 1029–42.

42 Neaves, W., 'The status of the human embryo in various religions', *Development* 144 (2017), pp. 2541–3.

43 To be clear, this is not a slight against religion. There are countless other examples where cultural views have hugely influenced scientific and medical opinion. To take another example, Guy Leschziner, in his book *The Nocturnal Brain: Nightmares, Neuroscience and the Secret World of Sleep* (Simon and Schuster, 2019), recounts how the emotional problem of hysteria was once thought by a male-dominated medical profession to be a specific condition of women, casued by the shifting of the womb from its normal position.

44 Re-opening discussion over current rules and laws could feasibly have negative consequences for IVF therapy too. Arguably, it was easier to debate moral issues respectfully in the 1980s compared to today. In an interview with *Nature Biotechnology* in 2017, Warnock said, 'I am sincerely thankful that our committee was not engaging with the public in the days of Twitter and emails.'

45 Appleby, J. B. and Bredenoord, A.L., 'Should the 14-day rule for embryo research become the 28-day rule?' *EMBO Molecular Medicine* 10 (2018), e9437.

46 Yan, W., 'An interview with Magdalena Zernicka-Goetz', *Biology of Reproduction* 96 (2017), pp. 503–4.

47 Morris, S.A. *et al.*, 'Dynamics of anterior-posterior axis formation in the developing mouse embryo', *Nature Communications* 3 (2012), p. 673.

48 Zernicka-Goetz, M. and Highfield, R., *The Dance of Life: Symmetry, Cells and How We Become Human* (W. H. Allen, 2020).

49 Deglincerti, A, *et al.*, 'Self-organisation of the *in vitro* attached human embryo', *Nature* 533 (2016), pp. 251–4.

50 Interview with Ali Brivanlou, 24 June 2019.

51 Ibid.

52 Ibid.

53 Ibid.

54 Chronopoulou, E. and Harper, J. C., 'IVF culture media: past, present and future', *Human Reproduction Update* 21 (2015), pp. 39–55.

55 Sunde, A. *et al.*, 'Time to take human embryo culture seriously', *Human Reproduction* 31 (2016), pp. 2174–82.

56 Swain, J. E. *et al.*, 'Optimizing the culture environment and embryo manipulation to help maintain embryo developmental potential', *Fertility and Sterility* 105 (2016), pp. 571–87.

57 Khosravi, P. *et al.*, 'Deep learning enables robust assessment and selection of human blastocysts after *in vitro* fertilisation', *npj Digital Medicine* 2 (2019), 21.

58 In her book *Hello World: How to be Human in the Age of the Machine* (Doubleday, 2018), Hannah Fry relates a note of caution about machine-learning algorithms. If a machine learns its own route to solving a problem, it may not be easy to understand how it arrives at an answer. One image-recognition algorithm, for example, would claim that a fuzzy set of pixels was a car, until one pixel was changed and then it would claim it was a dog. It might be hard to be reliant on an algorithm which solves problems in a way we can't easily understand.

59 Capalbo, A. *et al.*, 'Implementing PGD/PGD-A in IVF clinics: considerations for the best laboratory approach and management', *Journal of Assisted Reproduction and Genetics* 33 (2016), pp. 1279–86.

60 A list of genetic variations which can be screened by PGD, as approved by the UK's regulatory authority, which works independently but on behalf of the Government, can be found here: https://www.hfea.gov.uk/pgd-conditions/

61 de Melo-Martín, I., 'The challenge for medical ethicists: Weighing pros and cons of advanced reproductive technologies to screen human embryos during IVF', *Human Embryos and Preimplantation Genetic Technologies* (eds Sills, E. S. and Palermo, G. D.) pp. 1–10 (Academic Press, 2019).

62 Davis, D. M., *The Compatibility Gene* (Penguin, UK; Oxford University Press, USA, 2013).

63 Solomon, A., *Far from the Tree: Parents, Children and the Search for Identity* (Chatto and Windus, 2013).

64 Interview with Paula Garfield, 20 June 2019.

65 Hinsliff, G. and McKie, R., 'This couple want a deaf child. Should we try to stop them?' *Observer*, 9 March 2008.

66 Interview with Paula Garfield, *op. cit.*

67 Savulescu, J., 'Education and debate: Deaf lesbians, "designer disability" and the future of medicine', *British Medical Journal* 325 (2002), pp. 771–3.

68 Interview with Paula Garfield, *op. cit.*

69 Savulescu, 'Education and debate ...', *op. cit.*

70 Lovell-Badge, R., 'CRISPR babies: a view from the centre of the storm', *Development* 146 (2019), dev175778.

71 Doudna, J. and Sternberg, S., *A Crack in Creation: The New Power to Control Evolution* (The Bodley Head, 2017).

72 Jennifer Doudna, 'Into the Future with CRISPR Technology', the 2019 Nierenberg Prize for Science in the Public Interest, recorded at Scripps, California, on 7 October 2019, available online here: https://www.youtube.com/watch?v=cUe-cOgpDDw

73 Doudna and Sternberg, *A Crack in Creation, op. cit.*

74 Abbott, A., 'The quiet revolutionary: How the co-discovery of CRISPR explosively changed Emmanuelle Charpentier's life', *Nature* 532 (2016), pp. 432–4.

75 Lander, E.S., 'The Heroes of CRISPR', *Cell* 164 (2016), pp. 18–28.

76 Liang, P. *et al.*, 'CRISPR/Cas9-mediated gene editing in human tripronuclear zygotes', *Protein and Cell* 6 (2015), pp. 363–72.

77 The clinical trial registry shows that an ethics committee at the Harmonicare Shenzhen women and children's hospital had approved research for the 'Evaluation of the safety and efficacy of gene editing with human embryo *CCR5* gene'. This was later changed to say: 'Been withdrawn with the reason of the original applicants cannot provide the individual participants data for reviewing Safety and validity evaluation of HIV immune gene *CCR5* gene editing in human embryos'.

78 Regalado, A., 'EXCLUSIVE: Chinese scientists are creating CRISPR babies', *MIT Technology Review*, 25 November 2019.

79 These five videos can be watched on YouTube here: https://www.youtube.com/channel/UCn_Elifynj3LrubPKHXecwQ

80 During a panel discussion at Aspen Ideas Festival, 22 June 2019, I asked Dr Duanqing Pei, a professor of stem cell biology and academic director at the Guangzhou Institutes of Biomedicine and Health, whether or not the gene-edited twins are verified to exist. He replied that we don't know yet, and that we had to wait for the outcome of a full investigation. The session is available online here: https://www.youtube.com/watch?v=NDvgS8J5Gx8

81 Lovell-Badge, 'CRISPR babies ...', *op. cit.*

82 Belluck, P., 'Gene-Edited Babies: What a Chinese Scientist Told an American Mentor', *New York Times*, 14 April 2019.

83 Cohen, J., '"I feel an obligation to be balanced." Noted biologist comes to defense of gene editing babies', *Science*. Online here: https://www.sciencemag.org/news/2018/2011/i-feel-obligation-be-balanced-noted-biologist-comes-defense-gene-editing-babies (28 November 2018).

84 This, along with other details of He's work, is discussed in 'The CRISPR gene-edited babies and the doctor who made them – what happened?' on *Science Friction, with Natasha Mitchell*, on Australian Broadcasting Corporation's Radio National, available online here: https://www.abc.net.au/radionational/programs/sciencefriction/crispr-babies-what-happened-next/11163052

85 Lovell-Badge, 'CRISPR babies …', *op. cit.*

86 Johnson, M. H. and Elder, K., 'The Oldham Notebooks: an analysis of the development of IVF 1969–1978. IV. Ethical aspects', *Reproductive Biomedicine and Society Online* 1 (2015), pp. 34–45.

87 Brown, L., Brown, J. and Freeman, S., *Our Miracle Called Louise* (Paddington Press, 1979).

88 Wie, D., 'Gene Scientist Fired by College as China Says He Broke the Law', *Bloomberg*, 22 January 2019.

89 Cyranoski, D., 'What CRISPR-baby prison sentences mean for research', *Nature* 577 (2020), pp. 154–5.

3. A Force for Healing

1 Herzenberg, L. A. and Herzenberg, L. A., 'Genetics, FACS, immunology, and redox: a tale of two lives intertwined', *Annual Review of Immunology* 22 (2004), pp. 1–31.

2 Herzenberg, L. A., Herzenberg, L. A. and Roederer, M., 'A conversation with Leonard and Leonore Herzenberg', *Annual Review of Physiology* 76 (2014), pp. 1–20.

3 Ibid.

4 Herzenberg, L.A., 'The more we learn'. Available online here: https://www.kyotoprize.org/wp/wp-content/uploads/2016/02/22kA_lct_EN.pdf. *Kyoto Prize acceptance speech* (2006).

5 Lee grew up near Brighton Beach, home to a relatively large Jewish immigrant community. Some details about life at Brighton Beach during the 1950s is online here: http://brooklynjewish.org/neighborhoods/brighton-beach/

6 Herzenberg and Herzenberg, 'Genetics, FACS …', *op. cit.*

7 Herzenberg, L. A. and Herzenberg, L. A., 'Our NIH years: a confluence of beginnings', *Journal of Biological Chemistry* 288 (2013), pp. 687–702.

8 Herzenberg and Herzenberg, 'Genetics, FACS …', *op. cit.*

9 Ibid.

10 Herzenberg, Herzenberg and Roederer, 'A conversation with Leonard and Leonore Herzenberg', *op. cit.*

11 Interview with Leonore Herzenberg, 26 July 2019.

12 Herzenberg, Herzenberg and Roederer, 'A conversation with Leonard and Leonore Herzenberg', *op. cit.*

13 Linus Pauling, who would go on to win two Nobel Prizes, one for science and one for peace, was involved in the Federation of American Scientists, along with many other scientific leaders at Caltech, especially to protest against McCarthyism.

14 Herzenberg, 'The more we learn', *op. cit.*

15 Herzenberg, Herzenberg and Roederer, 'A conversation with Leonard and Leonore Herzenberg', *op. cit.*

16 Discussion with Elizabeth Simpson, 3 September 2019.

17 Interview with Leonore Herzenberg, *op. cit.*

18 Lee Herzenberg was interviewed by Mary Harris for the National Public Radio programme, 'Only Human: A Birth That Launched The Search For A Down Syndrome Test', broadcast on 26 April 2016. Available online here: https://www.npr.org/sections/health-shots/2016/04/26/475637228/only-human-a-birth-that-launched-the-search-for-a-down-syndrome-test

19 Ibid.

20 Ibid.

21 Ibid.

22 Ashoor Al Mahri, G. and Nicolaides, K., 'Evolution in screening for Down syndrome', *Obstetrician and Gynaecologist* 21 (2019), pp. 51–7.

23 Herzenberg, L. A., Bianchi, D. W., Schroder, J., Cann, H. M. and Iverson, G. M., 'Fetal cells in the blood of pregnant women: detection and enrichment by fluorescence-activated cell sorting', *Proceedings of the National Academy of Sciences of the USA* 76 (1979), pp. 1453–5.

24 Fan, H. C., Blumenfeld, Y. J., Chitkara, U., Hudgins, L. and Quake, S. R., 'Noninvasive diagnosis of fetal aneuploidy by shotgun

sequencing DNA from maternal blood', *Proceedings of the National Academy of Sciences of the USA* 105 (2008), pp. 16266–71.

25 Van Dilla, M. A., Fulwyler, M. J. and Boone, I. U., 'Volume distribution and separation of normal human leucocytes', *Proceedings of the Society for Experimental Biology and Medicine* 125 (1967), pp. 367–70.

26 Herzenberg, L.A., 'The more we learn', *op. cit.*

27 Early flow cytometers used light from mercury arc or halogen lamps.

28 Lanier, L. L., 'Just the FACS', *Journal of Immunology* 193 (2014), pp. 2043–4.

29 Two engineers, Russ Hulett and William Bonner, based in Joshua Lederberg's lab, helped Len modify the Los Alamos plans and build the first version of the cell sorter.

30 Interview with Paul Norman, 22 July 2019.

31 Discussion with Elizabeth Simpson, *op. cit.*

32 Zborowski, M. and Herzog, E., *Life is with People: The Culture of the* Shtetl (Schocken, New York, 1962).

33 Interview with Leonore Herzenberg, *op. cit.*

34 Zborowski and Herzog, *Life is with People, op. cit.*

35 Herzenberg, Herzenberg and Roederer, 'A conversation with Leonard and Leonore Herzenberg', *op. cit.*

36 Sweet, R. G., 'High Frequency Recording with Electrostatically Deflected Ink Jets', *Review of Scientific Instruments* 36 (1965), pp. 131–6.

37 This built upon the work of the nineteenth-century physicist Félix Savart, who showed that a small jet of liquid would break up into a stream of droplets if the jet passed through a nozzle vibrated appropriately.

38 In some instruments, cells are already separated in droplets when they meet the laser beam, but often cells are hydrodynamically focused for interrogation and then split into droplets.

39 An expert in flow cytometry, Viki Male at Imperial College London, told me this about the electrical plates on a flow cytometer: 'The housing on a modern machine makes it pretty difficult (though not impossible) to electrocute yourself by touching the plates, but anyone who has trained on an old enough machine will have a story of accidentally touching a charged plate and being thrown halfway across the room. Everyone who has been

electrocuted in the line of duty in this way is weirdly proud of the fact.'

40 Hulett, H. R., Bonner, W. A., Barrett, J. and Herzenberg, L. A., 'Cell sorting: automated separation of mammalian cells as a function of intracellular fluorescence', *Science* 166 (1969), pp. 747–9.

41 The cost of the first cell sorter is quoted in an article from the Stanford Medicine News Center, 31 October 2013, 'Leonard Herzenberg, geneticist who developed key cell-sorting technology, dies', available online here: http://med.stanford.edu/news/all-news/2013/10/leonard-herzenberg-geneticist-who-developed-key-cell-sorting-technology-dies.html

42 Melamed, M. R., 'A brief history of flow cytometry and sorting', *Methods of Cell Biology* 63 (2001), pp. 3–17.

43 Kamentsky, L. A., Melamed, M. R. and Derman, H., 'Spectrophotometer: new instrument for ultrarapid cell analysis', *Science* 150 (1965), pp. 630–1.

44 Koenig, S. H., Brown, R. D., Kamentsky, L. A., Sedlis, A. and Melamed, M. R., 'Efficacy of a rapid-cell spectrophotometer in screening for cervical cancer', *Cancer* 21 (1968), pp. 1019–26.

45 Len describes doing this in an interview recorded in 1991. Available online here: http://www.cyto.purdue.edu/cdroms/cyto10a/media/video/Herzenberghow.html

46 Fulwyler, M. J., 'Electronic separation of biological cells by volume', *Science* 150 (1965), pp. 910–11.

47 Robinson, J. P., 'Mack Fulwyler in his own words', *Cytometry Part A* 67A (2005), pp. 61–7.

48 German scientist Wolfgang Göhde, at the University of Münster, designed the very first machine capable of counting labelled cells, but this instrument didn't sort cells apart.

49 Robinson, 'Mack Fulwyler in his own words', *op. cit.*

50 Herzenberg, 'The more we learn', *op. cit.*

51 Herzenberg, Herzenberg and Roederer, 'A conversation with Leonard and Leonore Herzenberg', *op. cit.*

52 Herzenberg and Herzenberg, 'Genetics, FACS …', *op. cit.*

53 Keating, P. and Cambrosio, A., *Biomedical Platforms: Realigning the Normal and the Pathological in Late-Twentieth-Century Medicine* (MIT Press, Cambridge, Massachusetts, 2003).

54 Herzenberg, L. A, *et al.*, 'The history and future of the fluorescence-activated cell sorter and flow cytometry: a view from Stanford', *Clinical Chemistry* 48 (2002), pp. 1819–27.

55 Herzenberg and Herzenberg, 'Genetics, FACS …', *op. cit.*

56 Estimates of the flow cytometry market do vary widely. An estimate of $3.7 billion in 2018 is taken from an analysis in 2019 by MarketsandMarkets™ Inc. Available online here: https://www.marketsandmarkets.com/PressReleases/flow-cytometry.asp

57 Herzenberg, Herzenberg and Roederer, 'A conversation with Leonard and Leonore Herzenberg', *op. cit.*

58 Shapiro, H. M., 'The evolution of cytometers', *Cytometry A* 58 (2004), pp. 13–20.

59 These were different kinds of antibody produced in animals and screened for their ability to label certain types of cell. They weren't produced in as precise a way as monoclonal antibodies are produced today, and what they marked exactly wasn't always clear.

60 In more detail, haemoglobin is made up from four protein chains, two α-globins and two β-globins. The α-globin chain is encoded in two genes, *HBA1* and *HBA2*, and the β-globin chain is encoded by a single gene, *HBB*. Each of these proteins is bound to an iron-containing molecule which can bind oxygen (called heme, which is itself made from a series of reactions involving many other proteins and genes). In this way, one haemoglobin complex can bind four oxygen molecules. As blood flows through our lungs where oxygen levels are high, oxygen is taken up by haemoglobin. Oxygen is then released elsewhere in the body, where levels are lower.

61 Herzenberg, Herzenberg and Roederer, 'A conversation with Leonard and Leonore Herzenberg', *op. cit.*

62 I discussed this previously, and in more detail, in my book *The Beautiful Cure* (The Bodley Head, 2018)

63 This involves chopping and shuffling antibody genes – itself a wonderful and complex process.

64 A nuance here is that this type of selection can also happen to some extent outside of the bone marrow.

65 In a bit more detail, a B cell displays at its surface a version of the antibody it can produce, called the B cell receptor. If this receptor locks onto its target, the B cell is stimulated and will multiply. Some of the daughter B cells become factories for production of the useful antibody. Other daughter B cells will mutate the antibody genes randomly and be tested in case an even better version of the antibody has been produced. This

process is called affinity maturation, and explains why an antibody immune response improves over time. Some of these B cells will also live in the body for a long time, allowing the body to respond rapidly if the same threat is encountered again. The details of this process are exceptionally important for understanding what constitutes long-lasting immunity, and for the design of vaccines.

66 Melchers, F., 'Georges Köhler (1946–95)', *Nature* 374 (1995), p. 498.

67 Kohler, G. and Milstein, C., 'Continuous cultures of fused cells secreting antibody of predefined specificity', *Nature* 256 (1975), pp. 495–7.

68 At a New Year's party, while Milstein was discussing antibodies with his wife and colleagues, it was Lee Herzenberg who suggested a name for the immortal antibody-producing cells, each a hybrid of a myeloma cell and a B cell: a 'hybridoma'. The name has stuck, and is well known to biology labs using antibodies – which is virtually all biology labs.

69 Springer, T. A., 'Cesar Milstein, the father of modern immunology', *Nature Immunology* 3 (2002), pp. 501–3.

70 Rajewsky, K., 'The advent and rise of monoclonal antibodies', *Nature* 575 (2019), pp. 47–9.

71 Grilo, A.L. and Mantalaris, A., 'The Increasingly Human and Profitable Monoclonal Antibody Market', *Trends in Biotechnology* 37 (2019), pp. 9–16.

72 Guise, G., 'Margaret Thatcher's influence on British science', *Notes and Records of the Royal Society of London* 68 (2014), pp. 301–9.

73 Marks, L. V., *The Lock and Key of Medicine: Monoclonal Antibodies and the Transformation of Healthcare* (Yale University Press, 2015).

74 Ibid.

75 Koprowski, H. and Croce, C., 'Hybridomas revisited', *Science* 210 (1980), p. 248.

76 Croce, C. M., 'Hilary Koprowski (1916–2013): Vaccine pioneer, art lover, and scientific leader', *Proceedings of the National Academy of Sciences of the USA* 110 (2013), p. 8757.

77 Springer, 'Cesar Milstein …', *op. cit.*

78 Harding, A., 'Profile: Sir Greg Winter; humaniser of antibodies', *Lancet* 368 (2006), p. S50.

79 Rabbitts, T. H., 'Cesar Milstein: October 8, 1927 – March 24, 2002', *Cell* 109 (2002), pp. 549–50.

80 Rada, C., Jarvis, J. M. and Milstein, C., 'AID-GFP chimeric protein increases hypermutation of Ig genes with no evidence of nuclear localisation', *Proceedings of the National Academy of Sciences of the USA* 99 (2002), pp. 7003–8.

81 Melchers, 'Georges Köhler ...', *op. cit.*

82 When George Köhler took up leadership of a Max Planck institute, he specifically requested that his contract should include the possibility of him retiring at the age of fifty and receiving his pension in full. This is documented in Eichmann's 2005 book, *Köhler's Invention* (Birkhäuser Verlag, Springer Science, Basel, Switzerland, 2005). It is not known if Köhler actually intended to go ahead with early retirement; Köhler's wife Claudia 'vehemently refused' to discuss this or any other personal issue with Eichmann.

83 Fritz Melchers, quoted in *Köhler's Invention*.

84 Herzenberg and Herzenberg, 'Genetics, FACS ...', *op. cit.*

85 This was discovered soon after HIV was identified, in 1984: Klatzmann, D. *et al.*, 'Tlymphocyte T4 molecule behaves as the receptor for human retrovirus LAV', *Nature* 312 (1984), pp. 767–8; Dalgleish, A. G. *et al.*, 'The CD4 (T4) antigen is an essential component of the receptor for the AIDS retrovirus', *Nature* 312 (1984), pp. 763–7.

86 Doitsh, G. and Greene, W. C., 'Dissecting how CD4 T cells are lost during HIV infection', *Cell Host & Microbe* 19 (2016), pp. 280–91.

87 Global HIV and AIDS statistics: 2019 fact sheet. Published by UNAIDS. Available online here: https://www.unaids.org/en/resources/fact-sheet (Accessed September 2019.)

88 Horowitz, A. *et al.*, 'Genetic and environmental determinants of human NK cell diversity revealed by mass cytometry', *Science Translational Medicine* 5 (2013), 208ra145.

89 Smith, S. L. *et al.*, 'Diversity of peripheral blood human NK cells identified by single-cell RNA sequencing', *Blood Advances* 4 (2020), pp. 1388–1406.

90 Horowitz *et al.*, 'Genetic and environmental determinants ...', *op. cit.*

91 Spitzer, M. H. and Nolan, G. P., 'Mass cytometry: single cells, many features', *Cell* 165 (2016), pp. 780–91.

92 Shalek, A. K. *et al.*, 'Single-cell transcriptomics reveals bimodality in expression and splicing in immune cells', *Nature* 498 (2013), pp. 236–40.

93   Interview with Moshe Biton, 17 December 2019.

94   Interview with Aviv Regev, 17 August 2020.

95   Montoro, D. T. *et al.*, 'A revised airway epithelial hierarchy includes CFTR-expressing ionocytes', *Nature* 560 (2018), pp. 319–324.

96   Interview with Moshe Biton, *op. cit.*

97   It wasn't that the team doubted their method worked; they had verified it with a prior study on all the cells present in the retina, for example.

98   Interview with Aviv Regev, *op. cit.*

99   Interview with Moshe Biton, *op. cit.*

100 Plasschaert, L. W. *et al.*, 'A single-cell atlas of the airway epithelium reveals the CFTR-rich pulmonary ionocyte', *Nature* 560 (2018), pp. 377–81.

101 Three papers, published at the same time, reported this discovery in 1989: Kerem, B. *et al.*, 'Identification of the cystic fibrosis gene: genetic analysis', *Science* 245 (1989), pp. 1073–80; Riordan, J. R. *et al.*, 'Identification of the cystic fibrosis gene: cloning and characterization of complementary DNA', *Science* 245 (1989), pp. 1066–73; and Rommens, J.M., *et al.* 'Identification of the cystic fibrosis gene: chromosome walking and jumping', *Science* 245 (1989), pp. 1059–65.

102 Travaglini, K. J. and Krasnow, M. A., 'Profile of an unknown airway cell', *Nature* 560 (2018), pp. 313–14.

103 Interview with Aviv Regev, *op. cit.*

104 Rozenblatt-Rosen, O., Stubbington, M. J. T., Regev, A. and Teichmann, S. A., 'The Human Cell Atlas: from vision to reality', *Nature* 550 (2017), pp. 451–3.

105 Interview with Aviv Regev, *op. cit.*

106 Hiby, S. E. *et al.*, 'Association of maternal killer-cell immunoglobulin-like receptors and parental HLA-C genotypes with recurrent miscarriage', *Human Reproduction* 23 (2008), pp. 972–6.

107 Vento-Tormo, R. *et al.*, 'Single-cell reconstruction of the early maternal-fetal interface in humans', *Nature* 563 (2018), pp. 347–53.

108 Davis, D. M. *The Compatibility Gene* (Penguin, UK; Oxford University Press, USA, 2013).

109 Colucci, F., 'The immunological code of pregnancy', *Science* 365 (2019), pp. 862–3.

110 Interview with Muzlifah Haniffa, 11 September 2019.
111 Discussion with Muzlifah Haniffa, 7 November 2019.
112 Interview with Jack Kreindler, 13 August 2020.

4. The Multi-coloured Brain

1 Rapport, R. L. *Nerve Endings: The Discovery of the Synapse* (W.W. Norton, New York, 2005).
2 Camillo Golgi, Nobel Lecture, delivered on 11 December 1906. Published in *Nobel Lectures, Physiology or Medicine 1901–21* (Elsevier, Amsterdam, 1967). Available online here: https://www.nobelprize.org/uploads/2018/06/golgi-lecture.pdf
3 Wacker, D. *et al.*, 'Crystal Structure of an LSD-Bound Human Serotonin Receptor', *Cell* 168 (2017), pp. 377–89.
4 Wang, S. *et al.*, 'Structure of the D2 dopamine receptor bound to the atypical antipsychotic drug risperidone', *Nature* 555 (2018), pp. 269–73.
5 A positron emission tomography (PET) scan can also see what's going on inside a human brain. PET scans have a wide range of applications in general. One way in which they are used to analyse brain activity is that a radioactive glucose tracer is imaged to highlight areas of the brain taking up glucose for energy. There tends to be better resolution for imaging brain activity with fMRI than with PET scans, but there are pros and cons to each technique, and sometimes both types of scan are used.
6 In detail, fMRI is quite complex. A magnetic field is used to align hydrogen protons in the body. Then, a radio wave is used to push the protons out of alignment. When they then relax back into alignment they emit a signal which the fMRI machine picks up. The signal strength depends on the surroundings of the hydrogen proton, and is different in oxygenated and deoxygenated blood. The differences are subtle, and the technique relies on statistical tests and computational analysis to make measurements. In 2009, a famous experiment used fMRI to study the brain activity of a completely dead salmon. The dead fish was shown a series of photographs of humans in social situations, and fMRI scans were analysed to reveal the resulting brain activity. The experiment showed that if fMRI scan results were not analysed properly, all sorts of spurious results can occur, even for a dead fish looking

at pictures of people. An excellent web resource on how MRI and fMRI scans work, hosted by Oxford University, with text by Hannah Devlin and including a short animation narrated by Ruby Wax, is available here: https://www.ndcn.ox.ac.uk/divisions/fmrib/what-is-fmri/introduction-to-fmri

7 McClure, S. M. *et al.*, 'Neural correlates of behavioral preference for culturally familiar drinks', *Neuron* 44 (2004), pp. 379–87.

8 Plassmann, H., O'Doherty, J., Shiv, B. and Rangel, A., 'Marketing actions can modulate neural representations of experienced pleasantness', *Proceedings of the National Academy of Sciences of the USA* 105 (2008), pp. 1050–4.

9 Another particularly striking example comes from an analysis of brain activity in chocolate lovers, who were asked to eat chocolate until they could no longer stand to eat any more. Different parts of the brain lit up in the beginning, when they were enjoying the chocolate, compared to at the end, while they forced themselves to keep going. This study, which used PET imaging to capture brain activity, is reported here: Small, D. M., Zatorre, R. J., Dagher, A., Evans, A. C. and Jones-Gotman, M., 'Changes in brain activity related to eating chocolate: from pleasure to aversion', *Brain* 124 (2001), pp. 1720–33.

10 There are any number of other examples. See: Sahakian, B. J. and Gottwald, J., *Sex, Lies and Brain Scans* (Oxford University Press, Oxford, 2017).

11 Azevedo, F. A. *et al.*, 'Equal numbers of neuronal and nonneuronal cells make the human brain an isometrically scaled-up primate brain', *Journal of Comparative Neurology* 513 (2009), pp. 532–41.

12 Ecker, J. R. *et al.*, 'The BRAIN initiative cell census consortium: lessons learned toward generating a comprehensive brain cell atlas', *Neuron* 96 (2017), pp. 542–57.

13 Allen, N. J. and Barres, B. A., 'Glia – more than just brain glue', *Nature* 457 (2009), pp. 675–7.

14 The inferior status of glial cells is emphasised by them being named after the Greek word for glue. But to take just one intriguing experiment: mice injected with human glial cells (derived from donated human foetuses) showed improved learning and memory. See Han, X. *et al.*, 'Forebrain engraftment by human glial progenitor cells enhances synaptic plasticity and learning in adult mice', *Cell Stem Cell* 12 (2013), pp. 342–53.

15 Interview with Jeff Lichtman, 9 October 2019.

16 Lichtman, J. W., Livet, J. and Sanes, J.R.A., 'Technicolour approach to the connectome', *Nature Reviews Neuroscience* 9 (2008), pp. 417–22.

17 Livet, J. *et al.*, 'Transgenic strategies for combinatorial expression of fluorescent proteins in the nervous system', *Nature* 450 (2007), pp. 56–62.

18 Matz, M.V. *et al.*, 'Fluorescent proteins from nonbioluminescent Anthozoa species', *Nature Biotechnology* 17 (1999), pp. 969–73.

19 It was a postdoc in Lichtman's lab, Jean Livet, who had many of the ideas here and was the first author on the first Brainbow paper.

20 Steenhuysen, J., '"Brainbow" paints mouse neurons in bright colors', *Reuters*, 31 October 2007. Available online here: https://www.reuters.com/article/us-brain-colors/brainbow-paints-mouse-neurons-in-bright-colors-idUSN3131568320071031

21 Interview with Jeff Lichtman, *op. cit.*

22 Ibid.

23 Weissman, T. A. and Pan, Y. A., 'Brainbow: new resources and emerging biological applications for multicolor genetic labeling and analysis', *Genetics* 199 (2015), pp. 293–306.

24 Jeff Lichtman said this in his talk 'Connectomics' at TEDxCaltech, recorded at California Institute of Technology, Pasadena, California, 18 January 2013. Available online here: http://www.tedxcaltech.com/content/jeff-lichtman

25 Sporns, O., Tononi, G. and Kotter, R., 'The human connectome: A structural description of the human brain', *PLOS Computational Biology* 1 (2005), e42.

26 Sporns, O., *Discovering the Human Connectome* (MIT Press, Cambridge, MA, USA, 2012).

27 Seung, S., *Connectome: How the Brain's Wiring Makes Us Who We Are* (Allen Lane, London, 2012).

28 Seung, S., 'I am my connectome', TEDGlobal 2010. Available online here: https://www.ted.com/talks/sebastian_seung

29 Blakemore, S.-J., *Inventing Ourselves: The Secret Life of the Teenage Brain* (Doubleday, London, 2018).

30 Interview with Matthew Cobb, 18 October 2019.

31 Interview with Jeff Lichtman, 9 October 2019.

32 Ibid.

33 Lakadamyali, M., Babcock, H., Bates, M., Zhuang, X. and Lichtman, J., '3D multicolor super-resolution imaging offers improved accuracy in neuron tracing', *PLOS One* 7 (2012), e30826.

34 Sanes, J. R., 'After Cajal: from black and white to colour', in *Portraits of the Mind: Visualizing the Brain from Antiquity to the 21st Century* 'ed. Schoonover, C.' (Abrams, New York, 2010).

35 Ford, A. and Peat, F. D., 'The role of language in science', *Foundations of Physics* 18 (1988), pp. 1233–42.

36 Bargmann, C., Denk, W. and Graybiel, A., 'The Kavli Prize winners'. Interview by Darran Yates, *Nature Reviews Neuroscience* 13 (2012), pp. 670–4.

37 Denk, W. and Horstmann, H., 'Serial block-face scanning electron microscopy to reconstruct three-dimensional tissue nanostructure', *PLOS Biology* 2 (2004), e329.

38 Although Denk didn't know it at the time, he later learnt that a similar idea earlier had been reported in 1981: Leighton, S. B. 'SEM images of block faces, cut by a miniature microtome within the SEM – a technical note', *Scanning Electron Microscopy* (1981), pp. 73–6.

39 In fact, how to best measure a knife's sharpness is its own small research field, because so many factors are involved. See, for example: Schuldt, S., Arnold, G., Kowalewski, J., Schneider, Y. and Rohm, H., 'Analysis of the sharpness of blades for food cutting', *Journal of Food Engineering* 188 (2016), pp. 13–20.

40 Helmstaedter, M. *et al.*, 'Connectomic reconstruction of the inner plexiform layer in the mouse retina', *Nature* 500 (2013), pp. 168–74.

41 Kim, J. S. *et al.*, 'Space-time wiring specificity supports direction selectivity in the retina', *Nature* 509 (2014), pp. 331–6.

42 Bae, J. A. *et al.*, 'Digital museum of retinal ganglion cells with dense anatomy and physiology', *Cell* 173 (2018), pp. 1293–1306, e1219.

43 Abbott, A., 'Crumb of mouse brain reconstructed in full detail', *Nature* 524 (2015), p. 17.

44 Jeff Lichtman mentions this in a talk, 'Can the Brain's Structure Reveal its Function?' given at the Marine Biology Labs, Woods Hole, USA, on 6 July 2018. Available online here: https://www.mbl.edu/friday-evening-lectures-2018/

45 Kasthuri, N. *et al.*, 'Saturated reconstruction of a volume of neocortex', *Cell* 162 (2015), pp. 648–61.

46 Jeff Lichtman's talk, 'Can the Brain's Structure Reveal its Function?', *op. cit.*

47 Email correspondence with Jeff Lichtman, 6 November 2019.

48 Abbott, A., 'Neuroscience: solving the brain', *Nature* 499 (2013), pp. 272–4.

49 World Science Festival: Q and A with Jeff Lichtman, streamed live on 11 April 2018. Available online here: https://www.youtube.com/watch?v=h14hcBrqGSg

50 Nichol Thomson cut slices of worm, each around 50 nanometres thick, and took pictures of them with an electron microscope. Computer technology was too primitive to analyse the images in any automated way and two scientists, John White and Eileen Southgate, worked through them manually. Because the task of analysing the images was so laborious, many of the images Thomson took have never been studied, even to this day. Thomson was a skilled electron microscopist, having previously worked as a technician for Lord Victor Rothschild, but Sydney Brenner recalls in his memoir (*My Life in Science*, as told to Lewis Wolpert, Science Archive Limited, 2001) that he had great trouble employing Thomson because he didn't have any formal higher education. 'This was in the days,' Brenner says, 'when people began to worry about qualifications, which I think is complete nonsense!'

51 White, J. G., Southgate, E., Thomson, J. N. and Brenner, S., 'The structure of the nervous system of the nematode' *Caenorhabditis elegans*', *Philosophical Transactions of the Royal Society of London. B, Biological Sciences* 314 (1986), pp. 1–340.

52 Cook, S. J. *et al.*, 'Whole-animal connectomes of both *Caenorhabditis elegans* sexes', *Nature* 571 (2019), pp. 63–71.

53 Goodman, M. B. and Sengupta, P., 'How *Caenorhabditis elegans* senses mechanical stress, temperature, and other physical stimuli', *Genetics* 212 (2019), pp. 25–51.

54 Bargmann, C. I. and Marder, E., 'From the connectome to brain function', *Nature Methods* 10 (2013), pp. 483–90.

55 Finding a way to represent a connectome is part and parcel of understanding it. What comes to mind is an allegory from the Argentine storyteller Jorge Luis Borges (which I have mentioned before, in another context, in my first book, *The Compatibility Gene*, Allen Lane, 2013). *There was once an Empire where the art of*

map-making was celebrated and revered. The Cartographers' Guild had the ultimate ambition of attaining a description of the Empire that was perfect. But such a thing – a point-by-point description of everything there was – only produced a map exactly the same size as the Empire itself. The work of the greatest minds culminated in something entirely useless. The perfect map was left discarded, and subsequent generations gave less importance to the Art of Cartography. Similarly, a complete and exact map of the brain will be as complex and impenetrable as the brain itself – until we find a way to represent how it really works.

56 Portman, D. S., 'Neural networks mapped in both sexes of the worm', *Nature* 571 (2019), pp. 40–2.

57 'Insights of the decade. Stepping away from the trees for a look at the forest: Introduction', *Science* 330 (2010), pp. 1612–3.

58 Hegemann, P. and Nagel, G., 'From channelrhodopsins to opto-genetics', *EMBO Molecular Medicine* 5 (2013), pp. 173–6.

59 Nagel, G. *et al.*, 'Channelrhodopsin-2, a directly light-gated cation-selective membrane channel', *Proceedings of the National Academy of the Sciences of the USA* 100 (2003), pp. 13940–5.

60 Crick, F., 'The impact of molecular biology on neuroscience', *Philosophical Transactions of the Royal Society, London B: Biological Sciences* 354 (1999), pp. 2021–5.

61 Zemelman, B.V., Lee, G. A., Ng, M. and Miesenbock, G., 'Selective photostimulation of genetically charged neurons', *Neuron* 33 (2002), pp. 15–22.

62 Boyden, E. S., 'A history of optogenetics: the development of tools for controlling brain circuits with light', *F1000 Biology Reports* 3 (2011), p. 11.

63 A page from Deisseroth's lab book from 1 July 2004 showed that he was testing several different types of light-switchable channel proteins, as well as using different tools to express them in neurons, to see what might work well.

64 Smith, K., 'Neuroscience: Method man', *Nature* 497 (2013), pp. 550–2.

65 'Insights of the decade', *op. cit.*

66 Boyden, E. S., Zhang, F., Bamberg, E., Nagel, G. and Deisseroth, K., 'Millisecond-timescale, genetically targeted optical control of neural activity', *Nature Neuroscience* 8 (2005), pp. 1263–8.

67 Deisseroth, K., 'Optogenetics: 10 years of microbial opsins in neuro-science', *Nature Neuroscience* 18 (2015), pp. 1213–25.

68 Deisseroth, K. *et al.*, 'Next-generation optical technologies for illu-minating genetically targeted brain circuits', *Journal of Neuroscience* 26 (2006), pp.10380–6.

69 Colapinto, J., 'Lighting the brain: Karl Deisseroth and the opto-genetics breakthrough', *New Yorker*, 18 May 2015.

70 'The Consummate Neuro-oncologist; a profile of Michelle Monje', Ludwig Cancer Research 2019 Research Highlights. Available online here: https://www.ludwigcancerresearch.org/success-story/ludwigs-annual-research-highlights-report/

71 Adkins, T., 'Curing The Uncurable: Meet Dr Michelle Monje, the researcher powering cures for deadly brain tumors', in *Alex's Lemonade Stand Foundation Blog* 18 July 2018. Available online here: https://www.alexslemonade.org/blog/2018/07/curing-uncurable-meet-dr-michelle-monje-researcher-powering-cures-deadly-brain-tumors.

72 You'll have your own view of whether or not this sort of treat-ment of a mouse is necessary, cruel or both. Needless to say, strict ethical approval is needed for such work, which was of course obtained and closely adhered to.

73 Adamantidis, A. R., Zhang, F., Aravanis, A. M., Deisseroth, K. and de Lecea, L., 'Neural substrates of awakening probed with optogenetic control of hypocretin neurons', *Nature* 450 (2007), pp. 420–4.

74 Chen, I., 'The Beam of Light That Flips a Switch That Turns on the Brain', *New York Times*, 14 August 2007.

75 Colapinto, J., 'Lighting the brain … ', *op. cit.*

76 Deisseroth K., 'Optogenetics, iBiology Science Stories', recorded September 2016, available online here: https://www.ibiology.org/neuroscience/optogenetics/

77 Bandelow, B. and Michaelis, S., 'Epidemiology of anxiety disorders in the 21st century', *Dialogues in Clinical Neuroscience* 17 (2015), pp. 327–35.

78 Ibid.

79 Perez, C. C., *Invisible Women: Exposing Data Bias in a World Designed for Men* (Chatto and Windus, London, 2019).

80 Tye, K. M. *et al.*, 'Amygdala circuitry mediating reversible and bidirectional control of anxiety', *Nature* 471 (2011), pp. 358–62.

81 Kim, S. Y. *et al.*, 'Diverging neural pathways assemble a behavioural state from separable features in anxiety', *Nature* 496 (2013), pp. 219–23.

82 Jennings, J. H. *et al.*, 'Distinct extended amygdala circuits for divergent motivational states', *Nature* 496 (2013), pp. 224–8.

83 Ungless, M. A., Whistler, J. L., Malenka, R. C. and Bonci, A., 'Single cocaine exposure *in vivo* induces long-term potentiation in dopamine neurons', *Nature* 411 (2001), pp. 583–7.

84 Chen, B.T. *et al.*, 'Rescuing cocaine-induced prefrontal cortex hypoactivity prevents compulsive cocaine seeking', *Nature* 496 (2013), pp. 359–62.

85 Terraneo, A. *et al.*, 'Transcranial magnetic stimulation of dorsolateral prefrontal cortex reduces cocaine use: A pilot study', *European Neuropsychopharmacology* 26 (2016), pp. 37–44.

86 Ekhtiari, H. *et al.*, 'Transcranial electrical and magnetic stimulation (tES and TMS) for addiction medicine: A consensus paper on the present state of the science and the road ahead', *Neuroscience & Biobehavioural Reviews* 104 (2019), pp. 118–140.

87 Ferenczi, E. and Deisseroth, K., 'Illuminating next-generation brain therapies', *Nature Neuroscience* 19 (2016), pp. 414–16.

88 Maher, B., 'Poll results: look who's doping', *Nature* 452 (2008), pp. 674–5.

5. The Others Within

1 In one instrument, for example, gene fragments are captured on an array of sequence-specific templates which can be analysed in parallel.

2 Rose, C., Parker, A., Jefferson, B. and Cartmell, E., 'The Characterization of Feces and Urine: A Review of the Literature to Inform Advanced Treatment Technology', *Critical Reviews in Environmental Science and Technology* 45 (2015), pp. 1827–79.

3 Aziz, R. K., 'A hundred-year-old insight into the gut microbiome!' *Gut Pathogens* 1 (2009), p. 21.

4 LeBlanc, J. G. *et al.*, 'Bacteria as vitamin suppliers to their host: a gut microbiota perspective', *Current Opinion in Biotechnology* 24 (2013), pp. 160–8.

5 Fredrik Bäckhed, 'CRC 1182 host-microbe Interviews, Normal Gut Microbiota in Metabolic Diseases: an interview by Thomas Bosch', 20 December 2018. Available here: https://www.youtube.com/watch?v=hVzH8XY326s

6 Bäckhed, F. *et al.*, 'The gut microbiota as an environmental factor that regulates fat storage', *Proceedings of the National Academy of the Sciences of the USA* 101 (2004), pp. 15718–23.

7 Ley, R. E. *et al.*, 'Obesity alters gut microbial ecology', *Proceedings of the National Academy of the Sciences of the USA* 102 (2005), pp. 11070–5.

8 Turnbaugh, P. J. *et al.*, 'An obesity-associated gut microbiome with increased capacity for energy harvest', *Nature* 444 (2006), pp. 1027–31.

9 Yong, E., *I Contain Multitudes: The Microbes Within Us and a Grander View of Life* (The Bodley Head, London, 2016).

10 Ley, R. E., Turnbaugh, P. J., Klein, S. and Gordon, J. I., 'Microbial ecology: human gut microbes associated with obesity', *Nature* 444 (2006), pp. 1022–3.

11 Goodrich, J. K. *et al.*, 'Human genetics shape the gut microbiome', *Cell* 159 (2014), pp. 789–99.

12 Lecture by Eran Elinav on 'Host Microbiome Interactions in Health and Disease', given at the Kiel Life Science annual retreat in Schleswig, Germany, 16 November 2017. Available online here: https://youtu.be/2sfPHdhXJoE

13 Interview with Eran Elinav, 18 February 2020.

14 Jenkins, D. J. *et al.*, 'Glycemic index of foods: a physiological basis for carbohydrate exchange', *American Journal of Clinical Nutrition* 34 (1981), pp. 362–6.

15 Spector, T., *The Diet Myth; The Real Science Behind What We Eat*, (Weidenfeld and Nicolson, London, 2015).

16 Collaboration, N.C.D.R.F., 'Trends in adult body-mass index in 200 countries from 1975 to 2014: a pooled analysis of 1698 population-based measurement studies with 19.2 million participants', *Lancet* 387 (2016), pp. 1377–96.

17 Webb, P. *et al.*, 'Hunger and malnutrition in the 21st century', *British Medical Journal* 361 (2018), k2238.

18 Collaboration, N.C.D.R.F., 'Worldwide trends in body-mass index, underweight, overweight, and obesity from 1975 to 2016: a pooled analysis of 2,416 population-based measurement studies in 128.9

million children, adolescents, and adults', *Lancet* 390 (2017), pp. 2627–42.

19   Gardner, C. D. *et al.*, 'Effect of Low-Fat vs Low-Carbohydrate Diet on 12-Month Weight Loss in Overweight Adults and the Association With Genotype Pattern or Insulin Secretion: The DIETFITS Randomized Clinical Trial', *JAMA* 319 (2018), pp. 667–79.

20   Segal, E., Elinav, E. and Adamson, E., *The Personalized Diet: The Pioneering Program to Lose Weight and Prevent Disease* (Grand Central Life and Style, New York, 2017).

21   Ibid.

22   Interview with Eran Elinav, *op. cit.*

23   Chiu, C.-J. *et al.*, 'Informing food choices and health outcomes by use of the dietary glycemic index', *Nutrition Reviews* 69 (2011), pp. 231–42.

24   Interview with Eran Elinav, *op. cit.*

25   Zeevi, D. *et al.*, 'Personalized nutrition by prediction of glycemic responses', *Cell* 163 (2015), pp. 1079–94.

26   Cha, A. E., 'This diet study upends everything we thought we knew about "healthy" food,' *Washington Post*, 20 November 2015.

27   Segal, Elinav and Adamson, *The Personalized Diet, op. cit.*

28   Ibid.

29   Lecture by Eran Elinav on 'Host Microbiome Interactions …', *op. cit.*

30   Eran Segal presented these details at the Future of Individualized Medicine conference held at Scripps Research Translational Institute, Ja Jolla, California, 14–15 March 2019. Available online here: https://youtu.be/S26fCwDeiy0

31   Zeevi *et al.*, 'Personalized Nutrition …', *op. cit.*

32   The good and bad diets had been designed either by experts looking at all the information recorded or by computer algorithm directly. Either way, results were similar. For diets designed by computer, ten out of twelve particpants benefited during their week of being on a good diet plan.

33   Zmora, N., Zeevi, D., Korem, T., Segal, E. and Elinav, E., 'Taking it personally: personalized utilization of the human microbiome in health and disease', *Cell Host Microbe* 19 (2016) pp. 12–20.

34   The BBC sent a TV crew to Israel to film Elinav and Segal. The presenter, Saleyha Ahsan, put herself through the same regimen

used for participants in the original research. She learnt that her personal good food list included avocado, croissant, yoghurt and granola, omelette, chocolate and ice cream. Bad foods for her included grapes, pizza, pasta, tomato soup, orange juice and sushi. Like those in the original research, her microbiome shifted its composition between her good and bad diet weeks. This programme aired on 27 January 2016, as Episode 4, series 4, of *Trust me, I'm a Doctor*. Clips from this episode, including interviews with Elinav and Segal, are available online here: https://www.bbc.co.uk/programmes/articles/2lw8qKp7NFf7N7mhbXmsY34/why-do-some-people-put-on-weight-and-not-others-and-can-we-change-it

35 Eran Segal's TEDx talk, 'What is the best diet for humans?' TEDxRuppin, 20 July 2016. Available online here: https://youtu.be/0z03xkwFbw4

36 They created a website and a phone app to help anyone who wants to do this, but when I checked in January 2020, and again in January 2021, signing up was no longer possible.

37 Elinav and Segal are scientific consultants for DayTwo, the company doing this.

38 Eckel, R. H., 'Role of glycemic index in the context of an overall heart-healthy diet', *JAMA* 312 (2014), pp. 2508–9.

39 Katz, D. L. and Meller, S., 'Can we say what diet is best for health?' *Annual Review of Public Health* 35 (2014), pp. 83–103.

40 Sacks, F. M. *et al.*, 'Effects of high vs low glycemic index of dietary carbohydrate on cardiovascular disease risk factors and insulin sensitivity: the OmniCarb randomized clinical trial', *JAMA* 312 (2014), pp. 2531–41.

41 Kolodziejczyk, A. A., Zheng, D. and Elinav, E., 'Diet-microbiota interactions and personalized nutrition', *Nature Reviews Microbiology* 17 (2019), pp. 742–53.

42 One study published in 2019 examined the bacteria found in stool from twenty people, living in the UK or Canada. They found a total of 273 bacterial species, of which 105 had never even been seen before. See: Forster, S. C. *et al.*, 'A human gut bacterial genome and culture collection for improved metagenomic analyses', *Nature Biotechnology* 37 (2019), pp. 186–92.

43 Moss, M., *Salt, sugar, fat: how the food giants hooked us* (WH Allen, London, 2013).

44 Ng, M. *et al.*, 'Smoking prevalence and cigarette consumption in 187 countries, 1980–2012', *JAMA* 311 (2014), pp. 183–92.

45 Rimmer, A., 'Don't scrap the sugar tax, doctors tell Johnson', *British Medical Journal* 367 (2019), p. 17051.

46 Schirmer, M., Garner, A., Vlamakis, H. and Xavier, R. J., 'Microbial genes and pathways in inflammatory bowel disease', *Nature Reviews Microbiology* 17 (2019), pp. 497–511.

47 Walter, J., Armet, A. M., Finlay, B. B. and Shanahan, F., 'Establishing or exaggerating causality for the gut microbiome: lessons from human microbiota-associated rodents', *Cell* 180 (2020), pp. 221–32.

48 Berer, K. *et al.*, 'Gut microbiota from multiple sclerosis patients enables spontaneous autoimmune encephalomyelitis in mice', *Proceedings of the National Academy of the Sciences of the USA* 114 (2017), pp. 10719–24.

49 Britton, G. J. *et al.*, 'Microbiotas from humans with inflammatory bowel disease alter the balance of gut Th17 and RORgammat(+) regulatory T cells and exacerbate colitis in mice', *Immunity* 50 (2019), pp. 212–24.

50 Strachan, D. P., 'Hay fever, hygiene, and household size', *British Medical Journal* 299 (1989), pp. 1259–60.

51 Rakoff-Nahoum, S., Paglino, J., Eslami-Varzaneh, F., Edberg, S. and Medzhitov, R., 'Recognition of commensal microflora by toll-like receptors is required for intestinal homeostasis', *Cell* 118 (2004), pp. 229–41.

52 A short-chain fatty acid is made up from one to six carbon atoms, with hydrogen atoms attached, and one end has a carboxyl group (-COOH).

53 Three studies published in 2013, conducted independently, found that gut microbes secrete short-chain fatty acids which promote the production and activity of a type of T cell called a regulatory T cell, which is involved in dampening or regulating other immune cells. These papers are: Arpaia, N. *et al.*, 'Metabolites produced by commensal bacteria promote peripheral regulatory T-cell generation', *Nature* 504 (2013), pp. 451–5; Atarashi, K. *et al.*, 'Treg induction by a rationally selected mixture of *Clostridia* strains from the human microbiota', *Nature* 500 (2013), pp. 232–6; Smith, P. M. *et al.*, 'The microbial metabolites, short-chain fatty

acids, regulate colonic Treg cell homeostasis', *Science* 341 (2013), pp. 569–73.

54 Trompette, A, *et al.*, 'Gut microbiota metabolism of dietary fiber influences allergic airway disease and hematopoiesis', *Nature Medicine* 20 (2014), pp. 159–66.

55 Bottcher, M. F., Nordin, E. K., Sandin, A., Midtvedt, T. and Bjorksten, B., 'Microflora-associated characteristics in faeces from allergic and nonallergic infants', *Clinical & Experimental Allergy* 30 (2000), pp. 1590–6.

56 Hall, I. C. and O'Toole, E., 'Intestinal flora in newborn infants with a description of a new pathogenic anaerobe, *Bacillus difficilis*', *American Journal of Diseases of Children* 49 (1935), pp. 390–402.

57 Kelly, C. P. and LaMont, J. T., '*Clostridium difficile* – more difficult than ever', *New England Journal of Medicine* 359 (2008), pp. 1932–40.

58 Interview with Elizabeth Mann, 22 January 2020.

59 Scott, N.A. *et al.*, 'Antibiotics induce sustained dysregulation of intestinal T cell immunity by perturbing macrophage homeostasis', *Science Translational Medicine* 10 (2018), eaao4755.

60 Khoruts, A. and Sadowsky, M. J., 'Understanding the mechanisms of faecal microbiota transplantation', *Nature Reviews Gastroenterology & Hepatology* 13 (2016), pp. 508–16.

61 Shi, Y. C. and Yang, Y. S., 'Fecal microbiota transplantation: Current status and challenges in China', *JGH Open* 2 (2018), pp. 114–16.

62 Eiseman, B., Silen, W., Bascom, G. S. and Kauvar, A. J., 'Fecal enema as an adjunct in the treatment of pseudomembranous enterocolitis', *Surgery* 44 (1958), pp. 854–9.

63 Khoruts, A., 'Fecal microbiota transplantation – early steps on a long journey ahead', *Gut Microbes* 8 (2017), pp. 199–204.

64 Ibid.

65 van Nood, E. *et al.*, 'Duodenal infusion of donor feces for recurrent *Clostridium difficile*', *New England Journal of Medicine* 368 (2013), pp. 407–15.

66 Hui, W., Li, T., Liu, W., Zhou, C. and Gao, F., 'Fecal microbiota transplantation for treatment of recurrent *C. difficile* infection: An updated randomized controlled trial meta-analysis', *PLOS One* 14 (2019), e0210016.

67 DeFilipp, Z. *et al.*, 'Drug-resistant *e. coli* bacteremia transmitted by fecal microbiota transplant', *New England Journal of Medicine* 381 (2019), pp. 2043–50.

68 Blaser, M. J., 'Fecal microbiota transplantation for dysbiosis – predictable risks', *New England Journal of Medicine* 381 (2019), pp. 2064–6.

69 Terveer, E. M. *et al.*, 'How to: establish and run a stool bank', *Clinical Microbiology and Infection* 23 (2017), pp. 924–30.

70 Lloyd-Price, J., Abu-Ali, G. and Huttenhower, C., 'The healthy human microbiome', *Genome Medicine* 8 (2016), 51.

71 Giles, E. M., D'Adamo, G. L. and Forster, S. C., 'The future of faecal transplants', *Nature Reviews Microbiology* 17 (2019), p. 719.

72 Jabr, F., 'Probiotics are no panacea', *Scientific American* 317 (2017), pp. 26–7.

73 Interview with Eran Elinav, *op. cit.*

74 Suez, J., Zmora, N., Segal, E. and Elinav, E., 'The pros, cons and many unknowns of probiotics', *Nature Medicine* 25 (2019), pp. 716–29.

75 Klein, S. L., 'Parasite manipulation of the proximate mechanisms that mediate social behavior in vertebrates', *Physiology and Behavior* 79 (2003), pp. 441–9.

76 Wong, A. C. *et al.*, 'Gut microbiota modifies olfactory-guided microbial preferences and foraging decisions in drosophila', *Current Biology* 27 (2017), pp. 2397–404.

77 Leitao-Goncalves, R. *et al.*, 'Commensal bacteria and essential amino acids control food choice behavior and reproduction', *PLOS Biology* 15 (2017), e2000862.

78 Yuval, B., 'Symbiosis: Gut Bacteria Manipulate Host Behaviour', *Current Biology* 27 (2017), R746–R747.

79 Valles-Colomer, M. *et al.*, 'The neuroactive potential of the human gut microbiota in quality of life and depression', *Nature Microbiology* 4 (2019), pp. 623–32.

80 Cryan, J. F. and Dinan, T. G., 'Mind-altering micro-organisms: the impact of the gut microbiota on brain and behaviour', *Nature Reviews Neuroscience* 13 (2012), pp. 701–12.

81 Johnson, K. V. and Foster, K. R., 'Why does the microbiome affect behaviour?' *Nature Reviews Microbiology* 16 (2018), pp. 647–55.

82 Anderson, S. C., Cryan, J. F. and Dinan, T., *The Psychobiotic Revolution: Mood, Food, and the New Science of the Gut-Brain Connection* (National Geographic, Washington, 2017).

6. Overarching Codes

1 Hood, L., 'A personal journey of discovery: developing technology and changing biology', *Annual Review of Analytical Chemistry* 1 (2008), pp. 1–43.

2 Timmerman, L., *Hood: Trailblazer of the Genomics Age* (Bandera Press, 2017).

3 Interview with William J. Dreyer by Shirley K. Cohen, 18 February–2 March 1999, in *Caltech Oral Histories* (https://resolver. caltech.edu/CaltechOH:OH_Dreyer_W).

4 Hood, L. E., 'My life and adventures integrating biology and technology', in *The Inamori Foundation: Kyoto Prizes and Inamori Grants*, Vol. 18, pp. 110–66 (The Inamori Foundation, Japan, 2004).

5 Ponomarenko, E. A. *et al.*, 'The size of the human proteome: the width and depth', *International Journal of Analytical Chemistry* 2016 (2016), 7436849.

6 Hewick, R. M., Hunkapiller, M. W., Hood, L. E. and Dreyer, W. J., 'A gas-liquid solid phase peptide and protein sequenator', *Journal of Biological Chemistry* 256 (1981), pp. 7990–7.

7 This brief description belies the troubleshooting and ingenuity needed to get this to really work. One crucial problem in sequencing a protein is that the sample needs to be pure. If a mixture of proteins is present, it's very hard to work out the sequence of one from another. When the team were analysing the prion protein, for example, one of the problems which kept coming up was that two or three amino acids seemed to be present in each position of the protein's sequence. At first, scientists assumed the sample of protein wasn't pure enough, and so they tried different methods of isolating it. Still, whatever they tried, the sequence still looked a mess. Eventually, one of scientists involved, Steve Kent, decided to organise the results in terms of how abundant each amino acid was in each position of the sequence. He wrote down the sequence of the most abundant amino acids, then the second-most abundant amino acids, and so on. Suddenly, everything was clear. When he moved the sequence of second-most abundant amino acids two places to the right, it perfectly matched the sequence of most abundant amino acids. In other words, position one turned out to be position three in a fraction of the sample, and so on. He realised that the sample did contain mostly one type of protein but, for some reason

during the isolation process, one end of it was nibbled. With this insight, the prion protein sequence was clear. (Stanley Prusiner recounts this story in his autobiography, *'Madness and Memory'*, and in his Nobel Prize lecture.)

8 Prusiner, S. B., Groth, D. F., Bolton, D. C., Kent, S. B. and Hood, L. E., 'Purification and structural studies of a major scrapie prion protein', *Cell* 38 (1984), pp. 127–34.

9 I recall hearing Stanley Prusiner give a seminar at Harvard University's Molecular and Cellular Biology Department (*c.* 1995). He presented evidence that disease could be caused by prion protein, but some Harvard professors weren't fully convinced. One argument was that the protein samples might be contaminated with untraceable amounts of genetic material which could be the actual cause of disease. Prusiner had to go to extraordinary lengths to prove that protein molecules alone really could be the basis of an infectious disease.

10 Prusiner, S. B., 'Prions', *Proceedings of the National Academy of the Sciences of the USA* 95 (1998), pp. 13363–83.

11 Scheckel, C. and Aguzzi, A., 'Prions, prionoids and protein misfolding disorders', *Nature Reviews Genetics* 19 (2018), pp. 405–18.

12 Estrin, J., 'Kodak's first digital moment, *New York Times,* 12 August 2015.

13 Prusiner, S. B., *Madness and Memory: The Discovery of Prions – a New Biological Principle of Disease* (Yale University Press, New Haven, 2014).

14 Hood, 'A personal journey of discovery ...', *op. cit.*

15 Hood, L., 'A personal view of molecular technology and how it has changed biology', *Journal of Proteome Research* 1 (2002), pp. 399–409.

16 In time, Applied Biosystems became part of other companies. It became part of Perkin-Elmer from 1993, and Life Technologies from 2008. Thermo Fisher Scientific acquired Life Technologies in 2014.

17 Miller, M. *et al.*, 'Structure of complex of synthetic HIV-1 protease with a substrate-based inhibitor at 2.3 A resolution', *Science* 246 (1989), pp. 1149–52.

18 Cohen, J., 'Protease inhibitors: a tale of two companies', *Science* 272 (1996), pp. 1882–3.

19 Hood, 'A personal journey of discovery ...', *op. cit.*

20 Ciotti, P., 'Fighting disease on the molecular front: leroy hood built a better gene machine and the world beat a path to his lab', *Los Angeles Times*, 20 October 1985.

21 Email correspondence with Leroy Hood, 2 November 2020.

22 Sanger, F., Nicklen, S. and Coulson, A. R., 'DNA sequencing with chain-terminating inhibitors', *Proceedings of the National Academy of the Sciences of the USA* 74 (1977), pp. 5463–7.

23 For this work, Fred Sanger won a share of the Nobel Prize for Chemistry in 1980. Amazingly, this was his second Nobel. He had already won a Nobel Prize for Chemistry in 1958, for his method of sequencing proteins, which he had applied to studying insulin, for example. At the time I write this, in 2021, only four people have ever won two Nobel Prizes.

24 Leroy Hood, 'Revolutionising Healthcare: Systems Biology and P4 Medicine', a talk given at University College, Dublin, 15 September 2016. Available online here: https://youtu.be/HlQcH3zgoVs

25 Interview with Lloyd M. Smith by David C. Brock and Richard Ulrych, 2 March, 2008, in *Chemical Heritage Foundation, Oral History Program* (New Orleans, Louisiana).

26 Smith, L. M. *et al.*, 'Fluorescence detection in automated DNA sequence analysis', *Nature* 321 (1986), pp. 674–9.

27 Matthews, J., 'Caltech's New DNA-Analysis Machine Expected to Speed Cancer Research', *Washington Post*, 12 June 1986.

28 Timmerman, *Hood, op. cit.*

29 Venter, J. C., *A Life Decoded* (Allen Lane, London, 2007).

30 Sinsheimer, R. L., 'The Santa Cruz Workshop – May 1985', *Genomics* 5 (1989), pp. 954–6.

31 Hood was sceptical about the Human Genome Project at first. For him, it wasn't about the feasibility of the project, but whether or not such a huge undertaking would be worthwhile. He changed his mind when the broad implications of the project were discussed in Santa Cruz.

32 Sulston, J. and Ferry, G., *The Common Thread: A Story of Science, Politics, Ethics and the Human Genome*, (Bantam Press, London, 2002).

33 Interview with Leroy Hood, 27 March 2020.

34 Ibid.

35 The 1000 Genomes Project Consortium, *et al.*, 'A global reference for human genetic variation', *Nature* 526 (2015), pp. 68–74.

36 Sinsheimer, 'The Santa Cruz Workshop …', *op. cit.*

37 Chen, J. *et al.*, 'Pervasive functional translation of noncanonical human open reading frames', *Science* 367 (2020), pp. 1140–6.

38 Hood, L. and Rowen, L., 'The Human Genome Project: big science transforms biology and medicine', *Genome Medicine* 5 (2013), p. 79.

39 Hood, 'Revolutionising Healthcare', *op. cit.*

40 Angier, N., 'Great 15-Year Project To Decipher Genes Stirs Opposition', *New York Times,* 5 June 1990.

41 Interview with Leroy Hood, 28 March 2020.

42 Roberts, L., 'Caltech deals with fraud allegations', *Science* 251 (1991), p. 1014.

43 Dietrich, B., 'Future Perfect – Thanks To Bill Gates' $12-Million Endowment, Scientist Leroy Hood Continues His Search For A New Genetic Destiny', *Seattle Times*, 9 February 1992.

44 Hood, 'A personal journey of discovery …', *op. cit.*

45 Ideker, T., Galitski, T. and Hood, L., 'A new approach to decoding life: systems biology', *Annual Review of Genomics and Human Genetics* 2 (2001), pp. 343–72.

46 Speaking to me in 2020, Hood said that, all in all, he had given around $60 million of his own money for research in biotechnology across several projects. Luke Timmerman's biography of Hood says that the Institute for Systems Biology was set up with $5 million of Hood's money, and he didn't take any salary himself for its first two years.

47 *Systems Biology: a vision for engineering and medicine*: report from the Academy of Medical Sciences and the Royal Academy of Engineering, February 2007.

48 King, A. and O'Sullivan, K., 'New "scientific wellness" strategy could cut chronic illnesses and save money', *Irish Times*, 5 September 2018.

49 Leroy Hood, speaking at Geek Wire Summit 2019, Seattle, 7–9 October 2019. 'Power Talk: Leroy Hood' is available online here: https://youtube/bWCwTQ2hXYw

50 Interview with Leroy Hood, *op. cit.*

51 Leroy Hood, speaking at Geek Wire Summit, *op. cit.*

52 Roberts, P., 'Closure of high-tech medical firm Arivale stuns patients: "I feel as if one of my arms was cut off"', *Seattle Times*, 26 April 2019.

53 Hou, Y. C. *et al.*, 'Precision medicine integrating whole-genome sequencing, comprehensive metabolomics, and advanced imaging', *Proceedings of the National Academy of the Sciences of the USA* 117 (2020), pp. 3053–62.

54 Plomin, R. and von Stumm, S., 'The new genetics of intelligence', *Nature Reviews Genetics* 19 (2018), pp. 148–59.

55 Jennifer Doudna: Q and A: 'Towards the end of genetic disease?' Interview by Katia Moskvitch for the World Economic Forum, 20 January 2015. Available online: https://www.weforum.org/agenda/2015/01/qa-towards-the-end-of-genetic-disease/

56 Jolie, A., 'My medical choice', *New York Times*, 14 May 2013.

57 Jolie Pitt, A., 'Diary of a Surgery,' *New York Times*, 24 March 2015.

58 Proctor, R. N., 'The history of the discovery of the cigarette–lung cancer link: evidentiary traditions, corporate denial, global toll', *Tobacco Control* 21 (2012), pp. 87–91.

59 Friend, S. H. *et al.*, 'A human DNA segment with properties of the gene that predisposes to retinoblastoma and osteosarcoma', *Nature* 323 (1986), pp. 643–6.

60 Cohen, J. G., Dryja, T. P., Davis, K. B., Diller, L. R. and Li, F. P., 'RB1 genetic testing as a clinical service: a follow-up study', *Medical and Pediatric Oncology* 37 (2001), pp. 372–8.

61 Kuchenbaecker, K. B. *et al.,* 'Risks of breast, ovarian, and contralateral breast cancer for BRCA1 and BRCA2 mutation carriers', *JAMA* 317 (2017), pp. 2402–16.

62 Skol, A. D., Sasaki, M. M. and Onel, K., 'The genetics of breast cancer risk in the post-genome era: thoughts on study design to move past BRCA and towards clinical relevance', *Breast Cancer Research* 18 (2016), 99.

63 Even this may have little consequence. There are two copies of every gene in every cell (apart from sperm and egg cells), one inherited from each parent. So one reason why a mutation may not have any effect is that there's another copy of the gene which can cover any loss. In the case of *BRCA1* and *BRCA2*, however, a fault in just one copy is enough to increase a person's chance of developing cancer.

64 Tomasetti, C. and Vogelstein, B., 'Cancer etiology. Variation in cancer risk among tissues can be explained by the number of stem cell divisions', *Science* 347 (2015), pp. 78–81.

65 Lynch, T. J. *et al.*, 'Activating mutations in the epidermal growth factor receptor underlying responsiveness of non-small-cell lung cancer to gefitinib', *New England Journal of Medicine* 350 (2004), pp. 2129–39.

66 Recondo, G., Facchinetti, F., Olaussen, K. A., Besse, B. and Friboulet, L., 'Making the first move in EGFR-driven or ALK-driven NSCLC: first-generation or next-generation TKI?' *Nature Reviews Clinical Oncology* 15 (2018), pp. 694–708.

67 Cieslik, M. and Chinnaiyan, A. M., 'Global genomics project unravels cancer's complexity at unprecedented scale', *Nature* 578 (2020), pp. 39–40.

68 Salvadores, M., Mas-Ponte, D. and Supek, F., 'Passenger mutations accurately classify human tumors', *PLOS Computational Biology* 15 (2019), e1006953.

69 Gerstung, M. *et al.*, 'The evolutionary history of 2,658 cancers', *Nature* 578 (2020), pp. 122–8.

70 Scilla, K. A. and Rolfo, C., 'The role of circulating tumor DNA in lung cancer: mutational analysis, diagnosis, and surveillance now and into the future', *Current Treatment Options in Oncology* 20 (2019), 61.

71 Heitzer, E., Haque, I. S., Roberts, C.E.S. and Speicher, M. R., 'Current and future perspectives of liquid biopsies in genomics-driven oncology', *Nature Reviews Genetics* 20 (2019), pp. 71–88.

72 Melo, S. A. *et al.*, 'Glypican-1 identifies cancer exosomes and detects early pancreatic cancer', *Nature* 523 (2015), pp. 177–82.

73 Sheridan, C., 'Exosome cancer diagnostic reaches market', *Nature Biotechnology* 34 (2016), pp. 359–60.

74 Kottke, T. *et al.*, 'Detecting and targeting tumor relapse by its resistance to innate effectors at early recurrence', *Nature Medicine* 19 (2013), pp. 1625–31.

75 Helmink, B. A., Khan, M.A.W., Hermann, A., Gopalakrishnan, V. and Wargo, J. A., 'The microbiome, cancer, and cancer therapy', *Nature Medicine* 25 (2019), pp. 377–88.

76 Gopalakrishnan, V., Helmink, B. A., Spencer, C. N., Reuben, A. and Wargo, J. A., 'The influence of the gut microbiome on cancer, immunity, and cancer immunotherapy', *Cancer Cell* 33 (2018), pp. 570–80.

77 Jobin, C., 'Precision medicine using microbiota', *Science* 359 (2018), pp. 32–4.

78 Tanoue, T. *et al.*, 'A defined commensal consortium elicits CD8 T cells and anti-cancer immunity', *Nature* 565 (2019), pp. 600–5.

79 Davis, D. M., *The Compatibility Gene* (Penguin, UK; Oxford University Press, 2013).

80 Mastoras, R. E. *et al.*, 'Touchscreen typing pattern analysis for remote detection of the depressive tendency', *Scientific Reports* 9 (2019), p. 13414.

81 Yurkovich, J. T., Tian, Q., Price, N. D. and Hood, L., 'A systems approach to clinical oncology uses deep phenotyping to deliver personalized care', *Nature Reviews Clinical Oncology* 17 (2020), pp. 183–94.

Index

bipolar 92
and dendrites 91, 92, 98, 100
identifying in Brainbow pictures 93–5
Purkinje 92
in roundworms 101–2 *see also* brain, the
neurotransmitters 89, 91, 133
New York Times 25, 60, 106, 142, 146–7
NIH *see* National Institutes of Health
Nobel Prizes
for Chemistry (1958) 201n23; (1962) 43;
(1980) 139, 201n23; (2008) 18, 19;
(2014) 24–5
in Physics (1964) 165n36
in Physiology or Medicine (1906) 88–9;
(1958) 65; (1984) 74; (1997) 137; (2007)
171n7
Nolan, Gary 82

Obama, President Barack 137
obesity 11, 116, 117, 118, 122
OpenBiome, Cambridge, MA 132
optogenetics 102–6, 107–10
organ transplants 156
osteoporosis 144
Oszmiana, Ania 28–9
oxygen 3, 73, 89

P4 medicine 143
pancreatic cancer 151
Parkinson's disease 92
'passenger mutations' 150
patents 77, 78
Patterson, George 21, 22, 166n46
PCR *see* polymerase chain reaction
Pei, Dr Duanqing 176n80
Pepys, Samuel 13, 33
Perutz, Max 43
PET (positron emission topography) scans
185n5, 186n9
PGD *see* pre-implantation genetic diagnosis
Pius IX, Pope 46
placenta, the 37, 38, 39, 41, 86, 126
planets, discovery of 2
polymerase chain reaction (PCR) 139
Porteus, Matthew 59
potassium dichromate 88
Prasher, Douglas 17, 18–19
pregnancy 39–40, 41, 42, 113
and the immune system 86–7, 126
and screening/tests for Down's
syndrome 55, 67 *see also* IVF;
miscarriages
pregnancy tests 76

pre-implantation genetic diagnosis (PGD)
52, 53
prion proteins 136–7, 200n9
probiotics 132
Prokhorov, Aleksandr 165n36
propionic acid 127
protease inhibitors 138
protein analysers 136–7
protein synthesisers 137–8
proteins 3, 6, 10, 30–32, 73–4, 80, 81–2, 136,
141, 149, 151
algae 103, 104, 105
analysing 136–7
CD4 79
EGFR 149
fluorescent 93–5, *see also* green
fluorescent protein (GFP)
prion 136–7, 200n9
and rabies 133
receptor 26, 89, 91
red 93, 94, 95
sequencing 136, 199n7, 201n23
at synapses 89
toxic 29 *see also* antibodies
Prusiner, Stanley 136–7, 200n9
psychobiotics 134
Purdy, Jean 42, 43, 60
Purkinje neurons 92

Quake, Stephen 60, 67

rabies 133
Ramón y Cajal, Santiago 88, 92
reagents 73, 75
Regev, Aviv 83, 85
Revlimid 28
rheumatoid arthritis 76
RNA sequencing, single-cell 82–3
roundworms, mapping 101–2

Saeed, Mezida 29
Sanes, Joshua 94
Sanger, Frederick 139
Sasson, Steven 137
Savart, Félix 179n37
Savulescu, Julian 54–5
schizophrenia 12, 92, 106
science/scientific knowledge 157–9
Science (journal) 24, 48, 102, 104, 105
Segal, Eran 118–23, 124
The Personalized Diet (with Elinav) 122
Segal, Keren 120
semen *see* sperm